Center for the Study of National Reconnaissance Classics

THE GAMBIT STORY

CENTER FOR THE STUDY OF
NATIONAL RECONNAISSANCE
CHANTILLY, VA

APRIL 2012

Foreword

This volume re-publishes *The Gambit Story* as part of the *Center for the Study of National Reconnaissance's (CSNR) Classics* series. The introductory information explains how this history of the Gambit program offers rich detail about program development, as well as a unique insight into the management of the National Reconnaissance Office's (NRO's) Air Force element from 1963 to 1984. Frederic Oder and his co-authors made liberal use of Robert Perry's earlier account of the Gambit program for much of their research. However, the history also gives attention to broader Intelligence Community (IC) interests with discussions of the intelligence requirements and the role of the IC in Gambit operations. The authors drew on insight from senior Central Intelligence Agency (CIA) official, Roland Inlow, who had served as the chair of the Committee on Imagery Requirements and Exploitation (COMIREX) for the Director of Central Intelligence. Former Air Force officer and senior CIA official, John Schadegg also provided insight into the role and interests of the IC.

The redactions in this volume include the removal of figures showing KH-8 imagery because the Gambit KH-8 primary film record remains classified as of this writing in March 2012. However the Director of the National Geospatial-Intelligence Agency (NGA) provided the NRO with selected imagery products representing the KH-8 system when the NRO turned over Gambit artifacts to the National Museum of the United States Air Force (NMUSAF). We included eight of these images in a supplemental section at the end of the history. The NGA Director also included a highly informative 1977 Eastman Kodak briefing about the Gambit program.

The *Center for the Study of National Reconnaissance Classics* is a series of occasional CSNR publications whose purpose is to inform our readers about classic issues from the past. The books and monographs in the series most typically are histories, but they also could address lessons-learned topics, the legacy recognition of people and programs, insights into historically significant artifacts, or tutorials on the discipline of national reconnaissance. We issue the publications in the series on both an *ad hoc* basis, or in connection with a significant event. We are issuing a Gambit-Hexagon collection of histories in response to Director of the NRO Bruce Carlson's decision in June 2011 to declassify the programs and his subsequent declassification announcement on 17 September 2011. The Historical Documentation and Research (HDR) Section of the CSNR selected five classic histories of the Gambit and Hexagon programs:

- *A History of Satellite Reconnaissance—The Perry Gambit & Hexagon Histories* (by R. L. Perry)

- *The Gambit Story* (by F. C. E. Oder, J. C. Fitzpatrick, & P. E. Worthman)

- *The Hexagon Story* (F. C. E. Oder, J. Fitzpatrick, & P. E. Worthman)

- *Hexagon Mapping Camera Program and Evolution* (M. Burnett)

- *A History of the Hexagon Program—The Perkin-Elmer Involvement* (by R. J. Chester)

On 21 January 2012, the CSNR published the first volume in the Gambit-Hexagon CSNR Classics series, *A History of Satellite Reconnaissance—The Perry Gambit & Hexagon Histories*. We did this in support of the ceremony that marked the NRO turning over a collection of Gambit and Hexagon artifacts to the NMUSAF and their exhibit opening of these artifacts to the public. The opening of this exhibit represented the largest collection of satellite reconnaissance artifacts ever assembled and put on public display. That exhibit can serve as a companion resource to those who read the histories in this CSNR Classics collection.

Each of these histories offers a different perspective on the programs; the Perry Gambit and Hexagon histories are from the viewpoint of a former Air Force historian at RAND writing in response to tasking from the then NRO Program A (Air Force program); the Oder, et. al. Gambit and Hexagon histories are from the viewpoint of authors with program experience working under the sponsorship of the Deputy Director of the NRO; the Burnett Hexagon mapping system history is from the viewpoint of the Hexagon program office working under the direction of two Air Force officers in the program and the NRO Program A Director; and the Chester Hexagon history is from the viewpoint of Perkin-Elmer, which was an associate contractor for the Hexagon program.

All of the authors researched and wrote their histories during what some observers might describe as the height of the Cold War, from 1964 to 1985. This influenced them to react to and focus heavily on the threat from the former Soviet Union and its allies. Also, all of the authors had at least some degree of first-hand knowledge

about these programs, and in many cases, they had first-hand experience working in the programs. This gives you a window into what it was like to be a participant-observer in the development and operation of these film-return satellite photoreconnaissance systems during the Cold War.

Dr. James D. Outzen, the NRO Senior Historian and Chief of the CSNR's HDR section, is the editor for the Gambit-Hexagon CSNR Classics series. Dr. Outzen selected the five histories for this CSNR Classics series from the NRO Records Center and CIA archives that collectively best retell the impressive Cold War story about these programs. He has prepared a brief preface and introduction for each history to provide context and explain its significance.

When you read the histories you will note that some information is missing. Even though the Director of the NRO authorized the declassification of almost all the programmatic information about these programs, some information, because of its potential impact on other sources and methods, remains classified. Dr. Outzen usually let the redacted text stand on its own, but in some instances he has done some editing for readability. For some of the histories, Dr. Outzen has incorporated supplemental reference material into the publication.

Robert A. McDonald, Ph.D.

Director
Center for the Study of National Reconnaissance

Preface

Coinciding with the commemoration of the 50th Anniversary of the National Reconnaissance Office (NRO), the Director of the NRO, Mr. Bruce A. Carlson, publicly announced the declassification of the Gambit and Hexagon imagery satellite systems on 17 September 2011. This announcement constituted the NRO's single largest declassification effort in its history. The Gambit and Hexagon programs were active for nearly half of the organization's history by the time of the declassification announcement. Their history very much represents the NRO's history—one that is defined by supremely talented individuals seeking state of the art space technology to address difficult intelligence challenges.

The United States developed the Gambit and Hexagon programs to improve the nation's means for peering over the iron curtain that separated western democracies from east European and Asian communist countries. The inability to gain insight into vast "denied areas" required exceptional systems to understand threats posed by US adversaries. Corona was the first imagery satellite system to help see into those areas. It could cover large areas and allow the United States and trusted allies to identify targets of concern. Gambit would join Corona in 1963 by providing significantly improved resolution for understanding details of those targets. Corona provided search capability and Gambit provided surveillance capability, or the ability to monitor the finer details of the targets.

For many technologies that prove to be successful, success breeds a demand for more success. Once consumers of intelligence—analysts and policymakers alike—were exposed to Corona and Gambit imagery, they demanded more and better imagery. Consequently, the Air Force, who operated the Gambit system under the auspices of the NRO, entertained proposals for an improved Gambit system shortly after initial Gambit operations commenced. They received a proposal from Gambit's optical system developer, Eastman Kodak, for three additional generations of the Gambit system. Ultimately the Air Force settled on only developing the proposed third generation because the proposed second generation offered minimal incremental improvement and the fourth generation appeared technologically unachievable at the time. The third generation became known as Gambit-3 or Gambit-cubed while it was under development. Once it replaced the first generation, it simply became Gambit. The new Gambit system, with its KH-8 camera system, provided the United States outstanding imagery resolution and capability for verifying strategic arms agreements with the Soviet Union.

Corona was expected to serve the nation for approximately two years before being replaced by more sophisticated systems under development in the Air Force's Samos program. It turned out that Corona served the nation for 12 years before being replaced by Hexagon. Hexagon began as a Central Intelligence Agency (CIA) program with the first concepts proposed in 1964. The CIA's primary goal was to develop an imagery system with Corona like ability to image wide swaths of the earth, but with resolution equivalent to Gambit. Such a system would afford the United States even greater advantages monitoring the arms race that had developed with the nation's adversaries. The system that became Hexagon faced three major challenges. The first was development of the technology, which was eventually overcome by the Itek and Perkin-Elmer Corporations. The second was bureaucratic, deciding how the CIA and Air Force would cooperate in building such a system because they each had strengths and weaknesses in the development of national reconnaissance systems. The third challenge was to secure the resources that were required to build the most complicated and largest reconnaissance satellites at the time. By 1971, the NRO overcame the challenges to successfully launch the Hexagon satellite and fulfill, or even exceed, expectations for unparalleled insight into capabilities of US adversaries.

At the time of the Gambit and Hexagon declassification announcement, the NRO released a number of redacted Gambit and Hexagon documents and histories on its public website. One of the histories is contained in this volume.

The Gambit Story was written in 1988 by Frederic Oder, James Fitzpatrick, and Paul Worthman. Since its publication in 1991, *The Gambit Story* has served as a critical reference for the Gambit program, alongside the work of Robert Perry. Oder, Fitzpatrick, and Worthman each had varied and rich backgrounds in Air Force national reconnaissance programs that provided a strong foundation

for researching and writing the histories of satellite imagery programs. They were asked by then NRO Deputy Director, Jimmie D. Hill, to write individual histories of the Corona, Gambit, and Hexagon systems. All three have since preserved the essential history of the programs.

The Gambit Story is very rich in detail. The authors carefully document the origins of the first generation Gambit system and its KH-7 camera system, as well as the follow-on Gambit-3 system with its KH-8 camera system. The authors include a wide range of summary tables and information including details of each launch, companies and personnel involved in the launches, photographs and illustrations, and the capabilities of the systems. The history is well-documented and sourced.

Since the authors' backgrounds are in national reconnaissance programs—and primarily in the Air Force element of the NRO—they offer unique insight into the decisionmaking process for developing, launching, and operating national reconnaissance systems. Their Air Force perspective reveals valuable historical viewpoints that help document the Air Force element of the NRO's contributions through the Gambit systems to the nation's defense.

The Gambit Story joins five other volumes of Gambit and Hexagon histories that the Center for the Study of National Reconnaissance is reprinting in conjunction with the program declassifications. Those other volumes include *The Hexagon Story* also written by Oder, Fitzpatrick, and Worthman, Robert Perry's histories of Gambit and Hexagon, a history of the Hexagon mapping camera, a Perkin-Elmer history of Hexagon, and a compendium of key Gambit and Hexagon program documents. In total, this collection of Gambit and Hexagon publications provides the public with broad insight into previously classified programs. The volumes complement each other in providing details not found exclusively in any single program history volume.

At the time of this writing, KH-8 camera system imagery has not been declassified. I have included in a separate section of this publication a small number of KH-8 images that were released in conjunction with the Gambit declassification.

I have also included two additional Gambit documents in this publication. The first is a briefing that Eastman Kodak, the developers of the KH-7 and KH-8 camera systems, provided on the Gambit programs. The briefing is rich in detail and includes elegant drawings associated with the programs to help readers more fully understand the technical capabilities of the Gambit systems.

Additionally, I have included what became informally known as the "yellow brick road" pictures. These photographs show the sequence of assembling the Gambit flight vehicle in preparation for launch. The photographs reveal the number of steps necessary to successfully prepare each vehicle for successful launch.

Together, I hope these additional sections, along with *The Gambit Story*, will provide readers with more insight into the marvels of intelligence collection that the Gambit systems became.

I have chosen not to reprint pages that were redacted in their entirety in *The Gambit Story*. Those pages are: 95, 125, 137, 140 – 142, 144, 145, 147, 149 – 152, 166, and 169 – 172. We also did not reprint blank pages, which consist of pages 54, 102, 106, and 176. The unedited redacted *Gambit Story* can be found in the declassified records section of NRO.gov for those interested in reviewing a document with the completely redacted and blank pages.

The Gambit and Hexagon systems became reliable means for addressing difficult intelligence challenges once they became operational. The Gambit systems, in particular, provided high resolution imagery that was essential for understanding the strategic technical capabilities of the Soviet Union and other Cold War adversaries. These national reconnaissance systems dutifully provided the nation reliable vigilance from above until the next generation of imagery satellites advanced US intelligence collection capabilities.

James D. Outzen, Ph.D.

Chief, Historical Documentation and Research
The Center for the Study of National Reconnaissance

Table of Contents

The Gambit Story

Gambit Program - Eastman Kodak Co. 1977 Presentation

Gambit (KH-8) Imagery - 1966-1984

Gambit Program - Yellow Brick Road Presentation

Center for the Study of National Reconnaissance

The Center for the Study of National Reconnaissance (CSNR) is an independent National Reconnaissance Office (NRO) research body reporting to the NRO Deputy Director, Business Plans and Operations. Its primary objective is to ensure that the NRO leadership has the analytic framework and historical context to make effective policy and programmatic decisions. The CSNR accomplishes its mission by promoting the study, dialogue, and understanding of the discipline, practice, and history of national reconnaissance. The CSNR studies the past, analyzes the present, and searches for lessons-learned.

NRO APPROVED FOR
RELEASE 17 September 2011

~~Secret~~

~~NOFORN-ORCON~~
Handle via
BYEMAN-TALENT-KEYHOLE
Control Systems Jointly

The GAMBIT Story

~~Secret~~

BYE 140002-90
June 1991

Copy 8 of 25

WARNING NOTICE
INTELLIGENCE SOURCES OR METHODS INVOLVED
(WNINTEL)

NATIONAL SECURITY INFORMATION
UNAUTHORIZED DISCLOSURE SUBJECT TO CRIMINAL SANCTIONS

DISSEMINATION CONTROL ABBREVIATIONS

NOFORN - NOT RELEASABLE TO FOREIGN NATIONALS
ORCON - DISSEMINATION AND EXTRACTION CONTROLLED
          BY ORIGINATOR

Contents

Section		Page
	Preface	vii
1	Surprise Attack: A Haunting Concern	1
	Early Responses to the Concern	3
	Aircraft Dashes—Too Shallow and Too Seldom	3
	'Open Skies'—Too Altruistic	3
	Balloon Reconnaissance—Too Random	4
	The U-2—Once Too Often	4
	Discoverer/CORONA—In Place in Space	5
	Samos and the National Reconnaissance Office	6
	NRO Security	14
	A New Satellite Reconnaissance Need	14
2	Origins of GAMBIT	15
	Blanket and Sunset Strip	15
	US Intelligence Board Requirements	16
	Search For A Home: WADD to SSD to SAFSP	17
	Sunset Strip Goes 'Black'	18
	GAMBIT Varietals: Program 307, Exemplar, Cue Ball	22
	GAMBIT System Characteristics	25
	Land Recovery Versus Aerial Recovery	29
	Payload Development at Eastman Kodak	30
	Orbital-Control Vehicle Development at General Electric	32
	GAMBIT's Final Home—SAFSP	34
	Hitch-Up, Roll-Joint, and Lifeboat	37
3	GAMBIT Operations: The Early Flight Program	41
	The Hard Times of 1964	42
	Changing the Guard	44
	The Philadelphia Story	45
	On Schedule or Over-Target?	49
	Out of the Valley	52
4	Origins of GAMBIT-3	55
	Kodak Proposals for an Advanced GAMBIT	55
	The NRO Selects GAMBIT-3	58
	GAMBIT-3 Developments at Lockheed and Kodak	61
5	GAMBIT-3 Flight Program	69
	GAMBIT-1 or GAMBIT-3? How Many?	70
	GAMBIT-3 and the Needs of the Intelligence Community	71
	GAMBIT-3 and the GAMBIT-1 Heritage	72
	GAMBIT-3 Operations—The Flight Program	74

Contents (continued)

	Exercising Program Priorities	77
	The Block-II Series	77
	The Dual-Platen Camera	84
	GAMBIT Films	85
	Film-Read-Out GAMBIT (FROG)	86
	The GAMBIT Award Program	93
	The NASA 'Gambit'	94
	Responding to the Unexpected	96
	Life-Extension Changes	97
	Dual-Mode GAMBIT	97
	GAMBIT Reaches Full Potential	99
6	GAMBIT Program Costs	103
	GAMBIT-1/KH-7	103
	GAMBIT-3/KH-8	103
	GAMBIT in Retrospect	104
7	Development Management: Styles in Program Control	107
	On-Time Delivery: A Look at the Record	107
	The On-Time Delivery Problem: Contributing Factors	108
	AFBMD: High-Level Response No. 1	110
	CORONA: A Second High-Level Response	110
	SAFSP: The Third High-Level Response	112
	A Summing-Up	113
8	From 'Haunting Concern' to Informed Response	115
	Presidential Commendation	117
	Commendation to the GAMBIT Program	117
	Appendix A	119
	GAMBIT—Key Contributor to National Security Intelligence	119
	National Intelligence Requirements Management	119
	GAMBIT Imagery Security Policy	122
	Anticipation of Success	123
	Requirements Definition Challenge	123
	COMIREX Automated Management System (CAMS)	126
	National Imagery Exploitation Responsibilities	128
	Film-Dissemination Responsibilities	128
	National Photographic Interpretation Center (NPIC)	128
	Development of the National Imagery Interpretability Rating Scale	130
	Weather Support	132
	GAMBIT Intelligence Utility	136
	Satisfaction of Major Intelligence Requirements	138
	Scientific and Technical Intelligence	154

Contents (continued)

Appendix B 177
    A CORONA Summary 177
Appendix C 179
    Leningrad and LANYARD: Search for
    the GRIFFON 179
        Bigger Spacecraft, New Booster, Roll-Joint
        Needed 180
        The P-Camera Experiment 181
References 183
Index 190

Illustrations

	Page
President Dwight D. Eisenhower	5
James R. Killian, Jr.	5
Gillette Management Procedures for Atlas, Titan, Thor	7
Pied Piper (Samos) Management Channels—1956	8
CORONA Management Channels	9
George B. Kistiakowsky	11
Edwin H. Land	11
NRO Director Joseph V. Charyk	12
USAF MGen John L. Martin, Jr.	12
Final Samos Organization—1960	13
USAF BGen Robert E. Greer	18
USAF Col Paul J. Heran	18
GAMBIT Orbital-Control Vehicle (OCV)—1961 (Artist's Concept)	24
Numerical Summary of GAMBIT-1 Payload[33]	26
Details of ▓▓▓▓▓ Construction	31
SAFSP Organization in 1963	37
NRO Director Brockway McMillan	39
USAF Col Quentin A. Riepe	39
GAMBIT-1 Flight Summary—1963	42
GAMBIT-1 Flight Summary—1964	43
GAMBIT-1 Flight Summary—Jan-Jun 1965	44
NRO Director Alexander H. Flax	44
USAF Col William G. King	44
SAFSP Organization in 1965	46
NRO Organization in 1965	46
GAMBIT-1 Flight Summary—1965-67	52
GAMBIT-3 Pointing and Stereo Capability	56
GAMBIT-3 Roll-Joint (Payload Adapter Section)	57
Numerical Summary of GAMBIT-3 Payload	62
Lockheed GAMBIT-3 Program Office Organization	65
GAMBIT-3 Agena Vehicle	66
GAMBIT-3 Launch Vehicle at Liftoff	68
GAMBIT-3 'Factory-To-Pad' Concept	73
GAMBIT-3 Flight Summary—1966-67	74
GAMBIT-3 Flight Summary—1967-69	75
Dual-Recovery Module	76
Air-Recovery of GAMBIT Capsule by C-130	77
GAMBIT-3 Flight Summary—1969-72	78
USAF Col ▓▓▓▓▓	79
USAF Col Lee Roberts	79
GAMBIT-3 Ascent and Orbital Events	80
GAMBIT-3 Flight Summary—1972-76	81

Illustrations (continued)

Photograph of Forward Section of GAMBIT Spacecraft 82
Schematic of GAMBIT Spacecraft 82
Photograph of Aft Section of GAMBIT Spacecraft 82
GAMBIT Reconnaissance System 83
NRO Director John L. McLucas 84
USAF BGen David D. Bradburn 84
Deputy Defense Secretary David Packard 87
███ Director Leslie Dirks 87
DDS&T Carl Duckett 89
DCI Richard Helms 89
Assistant Defense Secretary Gardner Tucker 89
Eugene Fubini 89
COMIREX Chairman Roland Inlow 91
Presidential Science Adviser Lee DuBridge 91
President Richard M. Nixon 92
National Security Adviser Henry Kissinger 92
GAMBIT Government-Contractor Relationships 93
USAF Col ███ 99
USAF Col ███ 99
GAMBIT-3 Flight Summary—1977-84 101
GAMBIT-3 Mission Life Growth 105
GAMBIT-3 Resolution Improvement 105
The Discoverer Program Office, Spring 1960 111
Commendation to the GAMBIT Program 117
COMIREX Organization Chart 120
Surveillance Target Requirements by Intelligence Problem Category & Geographic Area 125
Elements of the COMIREX Requirements Structure 126
CAMS National Imagery Community Network 127
CAMS Environment 127
Arthur C. Lundahl Director, NPIC, 1961–1980 129
Program 417 Weather Satellites 133
Mean Cloud-Free Areas of the World in January 134
Mean Cloud-Free Areas of the World in July 135
GAMBIT Surveillance Coverage—Mission No. 4354, 17 Aug 84 137
(A) Delta-Class Hull-Staging Area at Severodvinsk—20 May 1973 139
(B) Y-Class Submarine at Komsomolsk Shipyard—6 Sep 1969 140
(C) HEXAGON/KH-9 Imagery of Submarines at Severodvinsk—19 Sep 1976 141
(D) GAMBIT/KH-8 Imagery of Submarines at Severodvinsk—19 Sep 1976 141
(E) Inflatable Dummy Y-Class Decoys at Severodvinsk—Jun-Jul 1974 142

Illustrations (continued)

(A) Tyuratam Missile Test Center, Complex A—16 Mar 1968 143
(B) Space Booster at Tyuratam Missile Test Center, Complex J—
    19 Sep 1968 144
(C) Tyuratam Missile Test Center, Complex J1 and J2 Showing
    Blast Damage—3 Jul 1969 145
(D) Shuanchengtzu Missile Test Center, PRC, Complex A—
    29 May 1967 146
(E) Shuanchengtzu Missile Test Center, PRC, Complex B—
    13 Nov 1968 147
(A) Itatka ICBM SS-7 Soft Site—18 Apr 1968 148
(B) Yoshkar Ola SS-13 ICBM Complex—13 Aug 1968 149
(C) Kozelsk SS-8 Missile in Silo—27 Jun 70 150
(D) Camouflage Attempts at Kartaly ICBM Complex—26 Aug 1969 .. 151
(E) Dummy Type-IIID Launch Sites at Plesetsk MSC—30 Aug 1969 ... 152
Borisov Army Barracks—15 Aug 1968 153
Abalakovo Phased-Array Radar 154
Plesetsk ICBM Silos 155
Soviet Aircraft Carrier Construction at Nikolayev Shipyard 156
Delta-Class Submarine With Missile Tubes Open at Severodvinsk 157
Yevpatoriyo Deep-Space Radar-Tracking Facility 158
New Soviet AWACS Aircraft 159
Typhoon Submarine at Severodvinsk 160
Type SO-209 Film 161
Type SO-112 Film 161
Kaz-A BLACKJACK Bomber at Kazan Airframe Plant 162
A Tower of Golden Gate Bridge as Imaged by GAMBIT 163
(A) 'Caspian Sea Monster'—19 Mar 1968 164
(B) Drawing of 'Monster' from GAMBIT Imagery 165
(C) Last GAMBIT Photo of 'Monster'—11 Aug 1984 165
(A) Comparison of Operational and Dummy Sites at Kartaly—
    8 Jun 1968 166
(B) Infrared Netting at SS-20 Mobile Base at Gresk—12 Apr 1981 167
Model of a Soviet Type-IIIC ICBM Site 168
(A) 169
(B) 170
(A) 171
(B) 172
(C) 172
Berenzniki Chemical Combine 173
Yurya Rail-to-Rail Transfer Point With Infrared-Reflectant Netting 174
AMM/SAM Launch Complex at Tallinn 182

SECRET
NOFORN-ORCON

Preface

This is the second volume in the history of the National Reconnaissance Program. The first volume related the story of CORONA—the first successful program in applying space vehicles to overflight reconnaissance operations. In its several evolutionary versions, CORONA steadily improved its photographic surveys of denied areas (with final resolutions of six to ten feet), operating in what the Intelligence Community calls "search mode."

This volume is the story of a companion photographic satellite called GAMBIT, which was developed to perform at even better resolutions than CORONA and work against specified targets—an operation usually referred to as "surveillance mode." GAMBIT fulfilled this surveillance function from July 1963 to April 1984.

In preparing this account, we appreciated the availability of an earlier volume prepared by Robert Perry and published in 1974. We have made liberal use of Perry's material, his documentary references, and his analysis of influences and events at the National Reconnaissance Office's Special Projects Office during the first half of the GAMBIT "era."

We are also grateful to Maj. Gen. David Bradburn, Dr. Joseph V. Charyk, Capt. Frank Gorman, USN, Col. ███████ Brig. Gen. William G. King, Maj. Gen. John L. Martin, Jr., Col. ███████ and Col. Lee Roberts—all military principals in the GAMBIT program—for extended personal interviews; to Rudi Buschmann, Robert Powell, and Peter Ragusa, Lockheed Missiles and Space Company (LMSC) principals; to Tom Diosy and Leslie Mitchell, Eastman Kodak (EK) principals, for data on EK participation; to ███████ and his associates at the National Photographic Interpretation Center (NPIC) for support in selecting and interpreting historic examples of GAMBIT product; to Lt. Col. ███████ SAFSS, Capt. ███████ SAFSP, and Donald E. Welzenbach, CIA, for essential assistance with sources and editing; to Roland Inlow, former chairman of COMIREX, for an overview of intelligence requirements; to ███████ former chairman of the Imagery Collection Requirements Subcommittee, for his contribution regarding the role of the Intelligence Community in GAMBIT operations; and to the legendary Arthur C. Lundahl for recollections presented in the final chapter.

The need for this series of histories was first envisioned by Jimmie D. Hill, Deputy Director, NRO. This volume, like the one on CORONA, was prepared under his sponsorship and constructive guidance.

August 1988
Sunnyvale, California

Frederic C.E. Oder
James C. Fitzpatrick
Paul E. Worthman

SECRET
Handle via
BYEMAN-TALENT-KEYHOLE
Control Systems Jointly
BYE 140002-90

Section 1

Surprise Attack: A Haunting Concern

The year was 1955. The President of the United States, dictating a letter to an old warrior-friend, was speaking with special intensity about a deep concern:

> Dear Winston:
>
> .... Your paper seems to me to under-emphasize a point of such moment that it constitutes almost a new element in warfare. I refer to the extraordinary increase in the value of tactical and strategic surprise, brought about by the enormous destructive power of the new weapons and the probability that they could be delivered over targets with little or no warning. Surprise has always been one of the most important factors in achieving victory. And now, even as we contemplate the grim picture described in your memorandum, we gain only the glimmering of the paralysis that could be inflicted on an unready fighting force, or indeed upon a whole nation, by some sudden foray that would place a dozen or more of these terrible weapons accurately on target.[1]

The President's closest associates were well-acquainted with this concern: they had heard it expressed in various forms on numerous occasions. James R. Killian, Jr., president of the Massachusetts Institute of Technology (MIT) and first science adviser to a US President, was a man given to steady, measured prose; he referred, in his memoir, to "this fear [which] haunted Eisenhower throughout his presidency."[2]

Apprehension over surprise attack was a novel presidential reaction, even for a former Supreme Commander. In spite of a lifetime spent in military service—where the expression "surprise attack" was an instructional and tactical commonplace—previous experiences had suggested the merest glimmering of what Eisenhower now felt. West Point and wartime days had taught a catalog of defenses against attack, but nothing could have prepared him for the realities of nuclear surprise.

In the years since Eisenhower had graduated from the US Military Academy, even geography had changed—and changed almost as dramatically as the tools of warfare. In 1919, the European boundary of the old Russian Empire was a line stretching from the eastern Baltic to the Black Sea, with Finland, Estonia, Latvia, Lithuania, Poland, and the Balkans buffering Germany, Austria, Italy, and France. This steady-state picture of Europe held until the fall of 1939, when Soviet incursions into eastern Poland were followed by similar actions against Estonia, Latvia, and Lithuania.

World War II experiences gave Eisenhower a duple view of the Soviets. Initially, he had seen them as mortal enemies allied with Nazi Germany and Fascist Italy. Then, abruptly, the Soviets and Hitler parted company and the Red Army was transformed into an ally—helpful and reasonably punctual in supporting the West. Eisenhower even developed a special confidence in his Soviet counterpart, Marshall Georgiy K. Zhukov.[3]

At the conclusion of the war, Eisenhower observed with sadness the pell-mell demobilization of Allied forces, contrasting so strongly with Soviet determination to hold to a strength of 5-6 million men, 50,000 tanks, and 20,000 aircraft. He saw how easily the Soviets shifted their European presence to a new boundary—110 miles west of Berlin. With feelings of deep concern, he watched the *coup d'etat* in Czechoslovakia, the use of the Red Army to support communist regimes in eastern Europe, the communist pressures on Italy and Finland, and the shooting down of transport planes over Yugoslavia.

The Soviets conquered eastern Europe with almost magical swiftness. As for western Europe—it lay helpless. Eisenhower voiced regret that "the Soviet Union had no intention of continuing its [wartime] policy of friendship, even on the surface . . . ."[4]

In 1948, he left his postwar position as chief of staff of the US Army to become president of Columbia University. But the realities of the Soviet "threat" followed him into academia. It was an ominous event, in 1949, when the Soviets detonated their first nuclear weapon and the Central Intelligence Agency wrote its first "estimate" of the possibility of surprise attack against the United States.

War began in Korea in 1950, with a surprise attack which awakened smoldering memories of a Sunday morning at Pearl Harbor. More shocking surprise came with the information that Soviet technology had been able to parallel US efforts: the US test of a hydrogen bomb in November 1952 was echoed by a similar Soviet test in 1953. By this time, Eisenhower had returned to public life as President of the United States. He described the view from the White House:

> Two wars, with the United States deeply engaged in one and vitally concerned in the other, were raging in Eastern Asia; Iran seemed to be almost ready to fall into Communist hands; the NATO Alliance had yet found no positive way to mobilize into its defenses the latent strength of West Germany; Red China seemed increasingly bent on using force to advance its boundaries; Austria was still an occupied country, and Soviet intransigence was keeping it so . . . . Communism was striving to establish its first beachhead in the Americas by gaining control of Guatemala.[5]

Worst of all, in 1955 the Soviets compounded the "threat" by building an operational bomber, the Myacheslav M-4, or BISON, which was equivalent in capacity and range to the US Air Force's B-52. Every day thenceforth, American cities and installations would be under threat of nuclear surprise attack.

Winston Churchill had eloquently described an Iron Curtain which, as it descended around the USSR and its satellites, effectively hid Soviet activities from the eyes of former wartime partners. It appeared that as Soviet expansionism became increasingly aggressive, Soviet homeland activities were becoming increasingly secretive. The reassurance which a nation normally obtained from knowing, on a day-to-day basis, what another nation was doing was no longer available. And the "balance of knowing" tilted alarmingly as the Soviets continued to enjoy access to worldwide current events, even as they concealed their own activities. The Soviet security apparatus grew each year, rivalled only by China's in size and effectiveness.

Early Responses to the Concern

The "haunting concern" began to magnify as intelligence sources picked up tantalizing hints of a Soviet ballistic missile program. A contemporary witness, Walter W. Rostow, writes that it was

> . . . a time when responsible American officials were authentically frustrated and alarmed by our inability to penetrate the closed society of the U.S.S.R. and establish with reasonable precision the scale and momentum of the Soviet program to develop nuclear delivery capabilities that could mortally threaten Western Europe, Japan, and the United States.[6]

In the midst of almost suffocating uncertainty, one major American counter action appeared in 1954, as work began on the Atlas ICBM. But there were more specific actions addressed to the central problem: how to open windows into a closed society. In chronological order, these efforts utilized available aircraft, diplomatic ventures, lighter-than-air devices, specialized aircraft, and satellites.

Aircraft Dashes—Too Shallow and Too Seldom

During and after 1949, there was a definite step-up in "peripherals" flown against the Soviet Union. These were flights in which standard or specially-equipped aircraft made brief incursions into or along Soviet territory for purposes of visual/photographic observation or electronic surveillance. Even at their best, these sorties had inherent range and altitude limitations; as shallow ventures into denied areas, they were infrequent and very dangerous. When one compared the enormous dimensions of the "problem-area" to the coverage achieved by sporadic flights, the productive capacity seemed almost inconsequential.

'Open Skies'—Too Altruistic

In 1955, President Eisenhower decided to use the occasion of a Summit Conference, scheduled for Geneva in July, to make a proposal to the Soviets for a peaceful and perhaps enduring resolution to the "haunting concern." The proposal, called "Open Skies," suggested that the United States and the Soviet Union should:

- exchange comprehensive military "blueprints," describing every military installation, and

- permit each other to make aerial photographs of these installations on a regular basis.

Eisenhower was keenly disappointed when Soviet Premier Bulganin and Party First Secretary Khrushchev rejected his plan. His reaction was reflected in two important conclusions: first, he believed that truly definitive evidence of Soviet intention was finally at hand: "Khrushchev's own purpose was evident—at all costs to keep the USSR a closed society"[7]; second, he felt a call to action: "When the Soviets rejected Open Skies . . . I conceded that more intelligence about their war-making capabilities was a necessity."[8]

Balloon Reconnaissance—Too Random

The RAND Corporation had anticipated the concerns of the early 1950s in 1946, when it began studying the military intelligence problems which might be posed by a closed society. One of RAND's subsequent conclusions was that camera-carrying balloons might be used to overfly the USSR. The fact that the balloons could be produced quickly and inexpensively, could fly very high (above fighter aircraft ceilings), and would be unmanned made them a possible reconnaissance option.

With the commercial availability of polyethylene film, the RAND proposal received serious consideration, since the non-extensible characteristic of the film made it possible to fly balloons at pre-selected constant-pressure altitudes. Operationally, one could launch in western Europe, fly at very high altitudes (say, 60,000 to 90,000 feet), drift across the USSR in photographing mode, and recover in mid-air over the ocean, somewhere between Taiwan and Alaska. After Soviet rejection of the "Open Skies" proposal, Eisenhower, on 27 December 1955, authorized such a balloon project (called GENETRIX) to become operational. Flights began on 22 January 1956 and were continued until 24 February with 516 releases. The operation was discontinued because of vigorous Soviet objection. (The GENETRIX camera and aerial recovery system became important contributors to the satellite reconnaissance technology of the 1960s.)

The U-2—Once Too Often

Another reason for discontinuing the GENETRIX flights was the advent of the U-2 aircraft designed by Lockheed's Clarence L. "Kelly" Johnson, which began flight tests in August 1955 and first overflew Soviet territory on 4 July 1956. This aircraft's flight schedule and performance were followed closely by the President, since the U-2 could go directly to points of interest and photograph priority targets, such as strategic airfields, radar installations, and missile test sites and launching facilities. The U-2 was used sparingly, discreetly, and successfully until May Day, 1960, when Francis Gary Powers failed to complete the only attempted border-to-border flight from Pakistan to Norway. When the President decided to cancel additional aircraft overflights, the United States was, once again, "blind."

Discoverer/CORONA—In Place in Space

In 1954 and early 1955, Dr. James Killian chaired a presidentially-authorized Technological Capabilities Panel which prepared a report for Eisenhower on the theme, "Meeting the Threat of Surprise Attack." The report pressed strongly for the development of overhead reconnaissance systems and was an important factor in convincing Eisenhower to proceed with U-2 operations. Having played this card with limited success, it now seemed timely to encourage the development of a spacecraft for reconnaissance purposes, particularly since such a system could avoid the operational limitations of balloons and aircraft.

President Dwight D. EISENHOWER

James R. KILLIAN, Jr.

Early in 1958, the Air Force and CIA began work on a space system. In public, the satellite was known as Discoverer, and appeared to be dedicated to examining and reporting on the space environment. It looked like, and behaved like, a normal "discovering" precursor to later *bona fide* military spacecraft, such as attack alarm, observation, and communication systems. The data it produced would facilitate future spacecraft designs and operational choices. So much for the security cover. In private, the true name of Discoverer was CORONA and its main purpose was the overhead reconnaissance of denied areas. CORONA stood at the intersection of six earlier achievements:

- Thor, the first high-thrust US space booster, had made its maiden flight in September 1957 and was now available to furnish 165,000 pounds of thrust—sufficient to lift a photographic payload into orbit;

- The Agena spacecraft, which was to be joined to the Thor and would house and operate the photographic payload, had been in development since July 1956;

- Re-entry vehicles, capable of protecting a payload from very high temperature as it passed through the earth's atmosphere, had been developed successfully for the Atlas, Titan, and Thor ballistic missile programs;

- A global network for controlling orbiting satellites was under construction and would be in operation in 1958;

- A camera capable of operating in a space environment had been built in 1955 for GENETRIX; a more complex camera, called HYAC, had been built for the follow-on WS-461L balloon program in 1956. An improved HYAC model could be constructed on short notice for use in a satellite;

- The equipment and techniques for in-flight retrieval of photographic payloads had been tested in GENETRIX operations and were available for use in CORONA.

The first attempt to launch CORONA was made in January 1959, one year after inception of the program. All early test flights were entirely experimental; they were ventures into a new world where critical environmental data were not only unavailable but frequently beyond reasonable conjecture. The first completely successful flight did not occur until 18 August 1960, when Discoverer-XIV, also known as CORONA mission No. 9009, returned with 3,000 feet of film showing 1,650,000 square miles of Soviet countryside and identifying ground objects with resolutions ranging upward from 35 feet.

Over the next 12 years, CORONA spacecraft made 145 flights, and the system's reliability, versatility, and photographic resolution were improved steadily and CORONA became the "search workhorse" of the US Intelligence Community.[9]

Samos and the National Reconnaissance Office

The Discoverer/CORONA program had not come into being independently; actually, it had evolved from a much larger reconnaissance program called, variously, WS-117L, FEEDBACK, Pied Piper, and Sentry, which had been proposed to the Army Air Force, in 1946, by Project RAND (of the Douglas Aircraft Corporation). In 1951, RAND, continuing its original study, defined the technical characteristics of a reconnaissance satellite, designed for television transmission of photography from space to ground stations. In October 1955, the Air Force made Wright Air Development Center (WADC) the manager for such a system; the first development plan was prepared in April 1956.

SECRET
NOFORN-ORCON

The concept of a satellite-borne observation platform performing worldwide reconnaissance was certainly the most revolutionary idea circulating within the Air Force in the early post-World War II period. The booster required to place such a platform in orbit was not even under study, let alone development. But nine years later, in 1955, with the first stirrings of intercontinental ballistic missile development, it became credible to consider space ventures: Atlas, Titan, and Thor, once satisfactorily tested, would each be capable of having their warheads replaced by space payloads. Thus, in 1955, RAND's revolutionary idea began to materialize as an on-going project.

Project activity at WADC was generally limited to studies and some experiments with components. Limitations would continue until there was a possibility of diverting one of the early missiles from its mandatory test program or from its swift progress to the Strategic Air Command's operational inventory. The highest priority in the United States had been given to the creation of an initial operational capability (IOC) with ICBMs; by comparison, the priority of WS-117L was much lower. But, by 1956, it did seem sensible to move the WS-117L study and planning activity closer to the booster program; soon the two became neighbors at the USAF's Western Development Division (WDD) in Inglewood, California.

There was a second, equally important, reason for the transfer: to place WS-117L near the quick-reaction management environment which enhanced the Atlas, Titan, and Thor developments. In September 1955, a unique management structure had been created by a committee advising the Secretary of the Air Force on the best way to streamline decision machinery for the ballistic missile program. The committee's recommendations were called "Gillette Procedures," after Hyde Gillette, the Air Force Deputy for Budget and Program Management. The radical nature of Gillette channels is contrasted with normal Air Force arrangements.

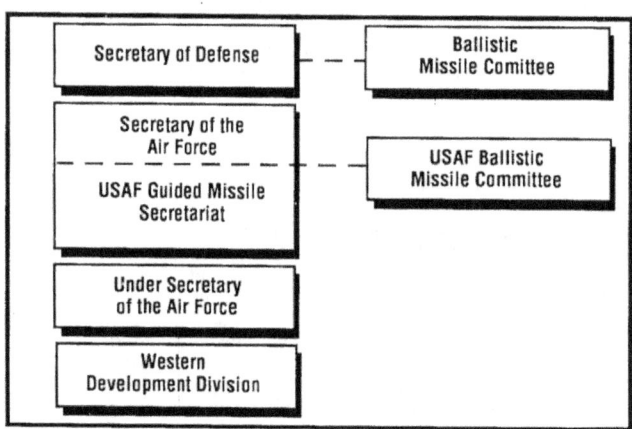

**Gillette Management Procedures
for Atlas, Titan, Thor**

SECRET

Handle via
BYEMAN-TALENT-KEYHOLE
Control Systems Jointly
BYE 140002-90

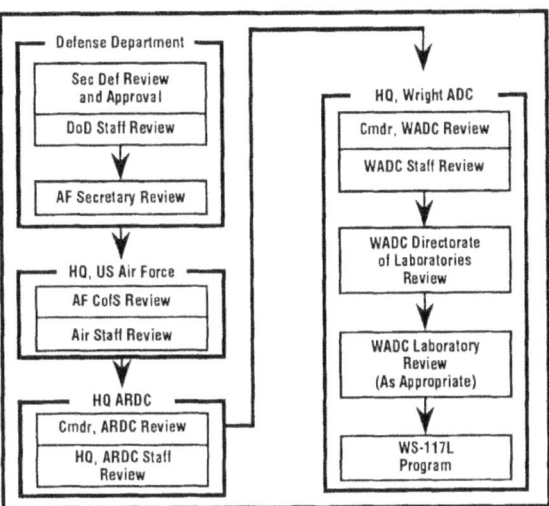

Pied Piper (Samos) Management Channels—1956

Despite the assignment of WS-117L to WDD, which placed it near the Gillette management process, the project was not legally within the ballistic missile ambit. The Air Force's intermediate military review echelons, smarting from exclusion by Gillette, were quick to point to this fact and to insist on their traditional mandate. In fact, the proximity of the ballistic missile program affected WS-117L adversely: it was hard for a budding enterprise to find a patch of sunshine amid towering ballistic giants. The WDD commander and his staff could not help comparing the priorities of the two programs: missile work was clearly of supreme importance and the bulk of energy should be devoted to it. Brig. Gen Bernard A. Schriever, WDD's commander, knew that if he paused for even a second on his missile mandate he would hear footsteps; his preoccupation with missiles was so exclusive that his public utterances did not even mention space systems until February 1957.

But changes were on the way. The strong national reaction to the Soviet's Sputnik flight, in October 1957, easily overshadowed the first successful Thor test flight the previous month. But both events combined to encourage the Air Force and the Central Intelligence Agency to break out a piece of WS-117L (Sentry was renamed Samos[10] in 1958) for a special purpose: the development of a quick-fix interim satellite reconnaissance program known as CORONA (treated earlier in this chapter).

The priority of CORONA was reflected in the fact that its management scheme out-Gilletted Gillette and was the ultimate any hardware development could hope to enjoy. All program management would be the responsibility of one person in the (entire) Air Force and one person in the CIA. Also,

Discoverer-CORONA would have the advantage of proximity to the Thor Office (for its booster), to the Agena Office (for its spacecraft), and to the Air Force's Ballistic Missile Division (AFBMD) satellite launching, tracking, control, and recovery facilities (for its operation).

Photographs produced on 18 August 1960, by the first successful CORONA flight, were impressive beyond hope and generated a surge of enthusiasm which spilled back into the Samos program. CORONA had never been intended as more than an interim quick-fix; now that success had been demonstrated, it was time to push Samos hard and achieve a truly sophisticated real-time-readout reconnaissance capability. Even the traditional limitation—availability of Atlas boosters for the heavier Samos payload—was becoming less of a problem.

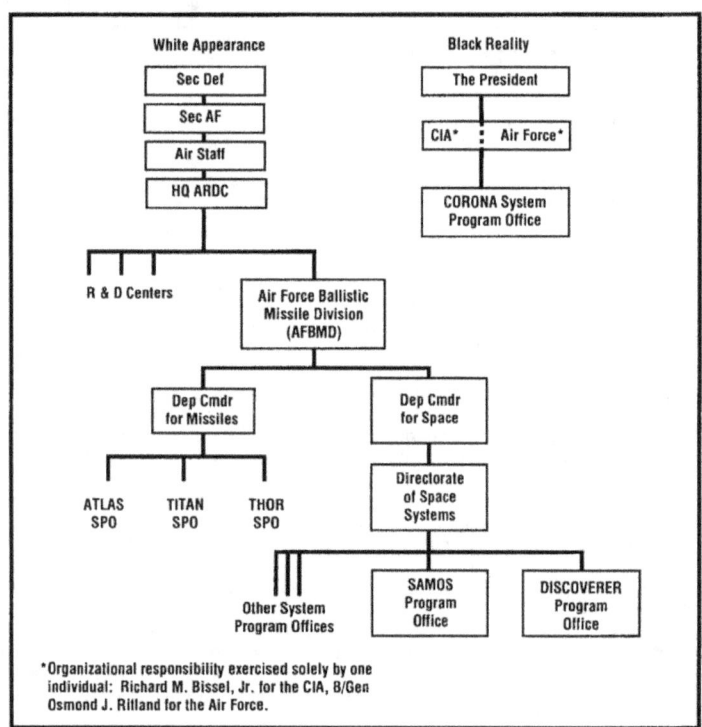

CORONA Management Channels

What would be the best means of encouraging and accelerating Samos? George B. Kistiakowsky, who had succeeded James Killian as science adviser to President Eisenhower, had been directed by the National Security Council to produce a "best means" plan. He had surveyed military space programs for some months and was not too impressed with what he found. In conversations with the USAF Air Research and Development Command's leadership, he had noted a strong aversion to "sacred cow" organizations like Schriever's Ballistic Missile Division (the new name for the Western Development Division) or for Gillette-type management channels. The threat to "command integrity" was abhorrent, said the "normal" Air Force, and once the ballistic program was "normalized," there should never again be any AFBMDs. Old-line research and development bureaucrats vowed that what had happened in ballistic missile management would not be permitted in the new field of space technology.

Kistiakowsky had also found that almost all ARDC Centers were viewing space projects—present and proposed—as plums to be coveted for that Center's future growth. Each Center dreamed of becoming the focus of Air Force space technology, encouraged in this brashness by popular anxiety over the Sputnik "space advantage" and rumors of new money to be poured into US programs. In the Air Staff itself, Kistiakowsky noted a preemptory announcement of the birth of an Astronautics Directorate (which the Air Force was forced to retract immediately). There was incessant, ill-advised discussion among newborn space enthusiasts about the value of space as a "high ground" and of the imperative need for a "cis-lunar defense capability." Kistiakowsky summed it up: "Frankly, it overwhelmed me. I still recall becoming indignant on discovering that the cost of exclusively paper studies in industrial establishments on 'Strategic Defense of Cis-Lunar Space' and similar topics amounted to more dollars than all the funds available to the National Science Foundation for the support of research in chemistry."[11]

Even in the more introspective and settled environment of the AFBMD, Kistiakowsky found a corporate opinion quite at odds with his own analysis of the present space reconnaissance need: "[T]hey believe that 'readout' Samos is much more promising than 'recovery' Samos."[12] In February 1960, he had taken a strong position on this subject: "[T]echnically the readout satellite is quite far in the future and, moreover, it has the inherent weakness of not providing sufficient detail of objects on the ground to be a useful instrument for our national security."[13] (Kistiakowsky recognized the desirability of a read-out satellite, but knew that existing data-link transmission technology was a severely limiting factor in readout capability. In short, this planned feature of Samos was ahead of its time; in a few years, however, it would become a viable part of reconnaissance technology.)

The U-2 shoot-down, on 1 May 1960, triggered a series of top-level decisions on Samos. The cancellation of aircraft overflight operations equated to the total loss of high-resolution observation of the USSR. Even if CORONA achieved success—and so far it had not—there would be an immediate need for much better resolution than it could provide; a system with the promise of Samos would continue to be absolutely essential.

On 26 May 1960, the President directed Kistiakowsky to set up a group to advise, as quickly as possible, on the best way to expand satellite reconnaissance options. Kistiakowsky chose some old friends to help frame the response: Killian now at MIT but still chairman of the President's Board of Consultants on Foreign Intelligence Activities (PBCFIA); Edwin H. Land of Polaroid, Carl Overhage, head of Lincoln Labs; and Richard M. Bissell, Jr., of

CIA. Since the CIA had no desire to expand its (limited) role in CORONA, Kistiakowsky pondered the capability of the Department of Defense to undertake a streamlined, augmented Samos. Managerially, he envisioned such a program as a super-CORONA. But without CIA involvement, could the DoD, or a military service, actually run a covert ("black") technical activity?[14]

Kistiakowsky was in close contact with Air Force Under Secretary Joseph V. Charyk, who echoed concerns over Samos, but argued strongly to keep the program in the Air Force. Charyk also insisted that, given a chance, he would prove that a program could be (both) in the Air Force and "black." Some

George B.
KISTIAKOWSKY

Edwin H.
LAND

months later, encouraged to "show how" this could be done, Charyk and Col. John L. Martin, Jr., invented a novel security strategy called "Raincoat." Raincoat was a security man's dream and a publicity man's nightmare. It proposed that the simplest way to hide a sensitive space program would be to sequester all military space programs—sensitive or no—from public view. Following the maxim that "at night all cats are gray," there would be no publicity release on any Air Force space program. Charyk discussed the concept in detail with Arthur Sylvester, Assistant Secretary of Defense for Public Information, who, after recovering from shock, actually became a supporter of the plan. It was important that the invention be dissociated from either Charyk or Sylvester, so the task of appearing to have generated the idea was assigned to Col. Paul E. Worthman, Chief, Plans and Programs Office, at

NRO Director Joseph V. CHARYK

USAF MGen John L. MARTIN, Jr.

the Space Systems Division (recently spun off from AFBMD and commanded by Maj. Gen. Osmond J. Ritland). Worthman's principal position at SSD was covert Air Force manager of CORONA. After a few briefings in appropriate Air Staff offices in the Pentagon, Worthman appeared in Charyk's office to make a final presentation to a large audience of hostile staffers, all of whom dreaded the thought of a broken rice bowl. At the conclusion of the briefing, Charyk approved his own invention and subsequently DoD Directive 5200.13 was issued, forbidding any publicity releases on Air Force space projects.

Kistiakowsky's Study Group made its recommendations to the President on 25 August 1960; they were approved the same day. In general, the Group proposed a fresh start for Samos, with a management structure closely modeled on the CORONA program. Procedures would be even more streamlined than those devised by Gillette for the ballistic missile program. The plan moved Samos out of the AFBMD environment, where it would have suffered from intense competition with ballistic missiles; out of the ARDC arena, where it could have been fought over by "space-hungry" Centers; out of the Air Staff, where it had been barely kept alive since 1956; and out of the Advanced Research Projects Agency (ARPA), where it had drifted aimlessly. The new organization was to be known overtly as the Office of Missile and Space Systems in the Office of the Chief of Staff of the Air Force.

Streamlining had finally been carried to the ultimate. The new Samos project office in Los Angeles would be housed in the same building as the new

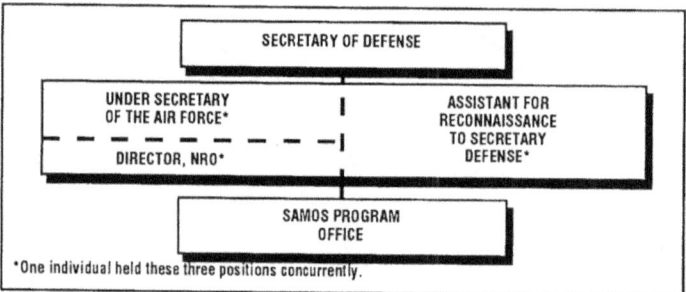

Final Samos Organization—1960

Space Systems Division. It would have direct access to all Air Force resources: Atlas boosters; Agena spacecraft; launching services at Vandenberg AFB; tracking and control services at Sunnyvale, California, and recovery services at Oahu.

In September 1961, the Department of Defense and the Central Intelligence Agency joined formally to create a National Reconnaissance Program (NRP) ("all satellite and overflight reconnaissance projects, whether overt or covert"). An office to manage the NRP, known overtly as the Office of Space Systems and covertly as the National Reconnaissance Office, was established in June 1962. The under secretary of the Air Force (then Charyk) was given additional duty as "the Special Assistant for Reconnaissance to the Secretary of Defense." Under this arrangement, the NRP would have easy access to Air Force space facilities and assets, while operating at the highest level of the Department of Defense.

For security purposes, the NRO was structured to look like "just another office" in the Air Force—possibly established to emphasize parochial interests in space. In reality, the NRO, from its earliest days, was an ecumenical, national effort, with representation from the entire intelligence community, including the three military services, the CIA, and the National Security Agency (NSA). Henceforth, the national reconnaissance needs for acquiring information over denied areas would be served by:

- a national requirements agency: the US Intelligence Board (USIB);

- a national reconnaissance agency: the National Reconnaissance Office (NRO); and

- a national interpretation agency: the National Photographic Interpretation Center (NPIC).

NRO Security

The NRO and its Program were concealed within and controlled by a special security system, designed by the CIA, known as BYEMAN. The existence of the NRO and all of the projects within the NRP were, and continue to be, highly protected, sensitive information.

A New Satellite Reconnaissance Need

Traditionally, experts in analyzing reconnaissance photography functionally divide it into two categories. One is called "search," and is dedicated to answering the question, "Is there something there?" CORONA's KH-4 panoramic camera was designed to photograph large contiguous areas in a single frame of film in order to provide answers to that question. Even though CORONA's resolution improved from its original 35–50 feet to 6–10 feet, its basic function remained search.

A second observation function is "surveillance." Surveillance is required after it has been decided that "There IS something of interest there," and says "I want to continue to watch that something, learn more about it, identify it, and classify it."

In most cases, *bona fide* surveillance was beyond CORONA's capability. The Intelligence Community soon expressed a need for a new satellite, which would sacrifice the extensive coverage capability of CORONA (millions of square miles) to acquire very detailed information on specified objects located in limited target areas (1–4 sq mi). The first successful satellite surveillance system was called GAMBIT, which carried a pointing or "spotting" camera with high-resolution capability.

The remainder of this volume is the story of GAMBIT's development, employment, maturation, and contributions to national security.

Section 2

Origins of GAMBIT

The definitive steps which led to the GAMBIT program were taken in early 1960—before the CORONA photographic reconnaissance satellite had its initial success. On 24 March 1960, the Eastman Kodak Company submitted an unsolicited proposal to the Air Force's Reconnaissance Laboratory at Wright Field. This proposal, which suggested development of a high-performance, 77-inch focal-length, catadioptric-lens camera suitable for satellite reconnaissance, had been developed by the Special Projects Group under Arthur B. Simmons, director of Research and Engineering in Kodak's Apparatus and Optical Division. At that time Kodak was under contract to the Reconnaissance Laboratory for development of a camera system for the OXCART aircraft program; the CIA was aware of the proposal because of its role in that program.

Blanket and Sunset Strip

On 17 June 1960, Kodak submitted another proposal, this time for a film-recoverable photographic reconnaissance system, which embodied a 36-inch lens camera to provide convergent-stereo area coverage of denied areas. Kodak called the system "Blanket" and claimed that, because of the planned use of existing technology, it could be made available in short order. This was followed, on 20 July 1960, by an elaboration of the 77-inch system proposed earlier, which used certain features of the Blanket concept and a film-handling technique proven feasible under the OXCART program. Kodak called the 77-inch system "Sunset Strip," both because of a then-popular TV program as well as the planned use of a strip camera.

Dr. Edwin H. Land, president of the Polaroid Corporation (and a very influential advisor to both Presidents Eisenhower and Kennedy) had, at that time, a close business relationship with Kodak as well as a personal and professional relationship with its key personnel, including Herman Waggershauser, vice president and general manager of the Apparatus and Optical Division, and Arthur Simmons. Simmons and Waggershauser showed the Sunset Strip proposal to Land, who, in mid-June, enthusiastically discussed the concept with Dr. Charyk. Later in June, at Charyk's request, Kodak sent him a copy of the Blanket proposal and a brief of the Sunset Strip concept. Kodak, concerned with keeping these ideas truly secret, used special CIA mail channels for correspondence with Charyk; few within the Eastman Kodak Company were aware of the proposals. On 5 July, Charyk and Simmons met to discuss both Blanket and Sunset Strip; this meeting reflected Simmons' growing enthusiasm for the 77-inch system's potential.[15]

US Intelligence Board Requirements

During the same period of 1959 and 1960, the US Intelligence Community was beginning to establish formal requirements for satellite collection capabilities; these would provide a firm basis for programs within the Samos effort. With the advent of the CORONA satellite program in late 1958, DCI Allen W. Dulles, with the concurrence of the US Intelligence Board (USIB) membership, established (in January 1959) a Satellite Intelligence Requirements Committee (SIRC). At a 1 June 1960 meeting, USIB agreed that the SIRC should develop an up-to-date statement of satellite intelligence requirements. (This action was a direct result of the downing of Gary Powers in a U-2 over the Soviet Union on 1 May 1960, which highlighted the need for a satellite-reconnaissance capability. The resulting SIRC report was submitted to the USIB in late June and was approved, with amendments, on 5 July 1960. The report, titled "Intelligence Requirements for Satellite-Reconnaissance Systems of which Samos is an Example," was sent to Secretary of Defense Neil McElroy. In a letter of transmittal, USIB Chairman Dulles stressed that the fulfillment of these requirements was critical to US security.[16]

The requirements outlined in the SIRC report called for a satellite reconnaissance system capable of obtaining coverage of denied areas at object resolutions of approximately 20, 5, ▓▓▓▓ on a side.

According to the SIRC document, the first and most urgent need was for a photographic search system capable of locating suspected ICBM launching sites in that part of the USSR covered by a railroad network. This would require a resolution approaching 20 feet on a side. A second priority requirement was to cover the same area with a resolution approaching five feet on a side, in order to obtain more descriptive information on the ICBM installations. The third priority was for a system which could provide a resolution better than five feet on a side, in order to supply, before the end of 1962, technical characteristics of the highest priority targets.

At a 5 July 1960 meeting, the USIB also concurred in the suggestion of DDCI Charles P. Cabell that the feasibility of consolidating the SIRC and the Ad Hoc Requirements Committee (ARC) be studied.[17] The resulting report recommended such a consolidation and, on 9 August 1960, the USIB approved DCID No. 2/7 establishing a Committee on Overhead Reconnaissance (COMOR)—comprising the ARC and the SIRC—to provide a focal point for information on, and the coordinated development of, foreign intelligence requirements for overhead reconnaissance operations over denied areas.

In addition to adapting priority objectives and requirements established by USIB, its members, or other committees, to the capabilities of existing and potential systems, COMOR was to examine and recommend dissemination procedures and special security controls required for operational guidance. COMOR was to consist of representatives of USIB agencies with a chairman designated by the DCI in consultation with, and the concurrence of, the USIB. James Q. Reber, who had been chairman of the ARC since December 1955, was the first chairman of COMOR; his deputy was Air Force Col. ▓▓▓▓

~~SECRET~~
~~NOFORN-ORCON~~

During the same month that COMOR was established, President Eisenhower, presiding over a special meeting of the National Security Council, directed the Air Force to give high priority to developing a film-return satellite system for providing high-resolution stereo photography (this became the basis for Project GAMBIT). At the same time, Eisenhower directed the Air Force to give the remaining Samos program a lower priority.

Search For A Home: WADD to SSD to SAFSP

Although Kodak was endeavoring to limit "need-to-know," Blanket and Sunset Strip were becoming known to a few people: some at Wright Field, some within the Samos organization of the Space Systems Division (SSD), and some within the Air Staff. The Air Staff decision to have Wright Field contract with Kodak for an engineering model of the 77-inch system soon became entangled in routine Air Force channels. Because of Charyk's interest, Simmons urged his Air Staff contact to handle the study through the Air Force space organization, rather than through the Wright Field Reconnaissance Laboratory.[18] On 13 August, the Air Force Staff rescinded its directive to Wright Air Development Division (WADD), redirecting the work to the Samos program. In forwarding copies of the earlier studies to the Space System Division on 13 August, Kodak proposed a 90-day Phase-I stage (design to mock-up) to cost ▓▓▓▓▓▓ and a subsequent Phase-II effort to include design, construction, test, and flight test of development models and prototype camera systems. Kodak noted the impossibility of projecting development costs until completion of the Phase-I activity and acknowledged the uncertainty of compatibility between the camera system and available boost, orbit, and recovery subsystems. Nevertheless, the contractor reaffirmed the feasibility of providing 2- to 3-foot ground resolution with a high-acuity, stereo-coverage, surveillance camera system placed in a short-lived satellite vehicle.[19]

Within 24 hours of receiving the Kodak studies and summary proposal, the Space Systems Division began processing a letter contract. About the same time, responsibility for the Samos program was transferred from the Space Systems Division to the newly-created Secretary of the Air Force Samos Project Office, which subsequently became the Secretary of the Air Force Special Projects Office (SAFSP). The office's military director, Brig. Gen. Robert E. Greer, had been transferred to Inglewood, California, from a previous assignment as the USAF's assistant chief of staff for guided missiles.

Under SAFSP direction, a competition was being held for the Samos E-6 program; proposals were to be submitted in October 1960. The E-6 project was a part of Samos (or Air Force Weapon System-117L) that had begun in 1960, long after the first General Operating Requirement had established WS-117L (in 1954). Samos originally had two planned photographic capabilities—"Pioneer" and "Advanced"—which were designated E-1 and E-2. These involved the on-orbit exposure and processing of film, translation of that imagery into an electrical signal by means of a flying-spot scanner, and transmission of the signal to earth for subsequent recomposition as a picture. The readout photographic versions of Samos were limited by state-of-the-art electronics to a 6-Megabit carrier—a limitation which, in 1958, caused priority

~~SECRET~~

Handle via
BYEMAN-TALENT-KEYHOLE
Control Systems Jointly
BYE 140002-90

to be given to film-recovery systems. E-3 was the designator for a system which substituted photo-sensitive electrostatic tape for film; E-4 was used to identify a proposed but unofficial mapping/geodetic photographic system; E-5 was a recoverable satellite with a large recovery vehicle; and E-6 was a recoverable-film search system with several times the capability of CORONA. E-1, E-2, and E-3 were readout systems, E-5 and E-6 were film-recovery systems. Only E-1, E-2, and E-6 ever flew—none with especial success. One of the E-6 camera competitors was Kodak and one of the spacecraft competitors was the General Electric Company, which was developing reentry vehicles for the CORONA program in its Chestnut Street Facility in Philadelphia, Pennsylvania.

USAF BGen Robert E. GREER

USAF Col Paul J. HERAN

Sunset Strip Goes 'Black'

On 10 September 1960, Charyk met with Greer, Col. Paul J. Heran (Chairman of the E-6 Source Selection Board), and Lt. Col. James Seay (Greer's procurement chief), to review proposed programs, including the Sunset Strip effort. The meeting resulted in a recommendation to Charyk to proceed with both E-6 (which had the potential of being twice as good as CORONA) and Sunset Strip. Charyk directed that Sunset Strip be developed and that this be done on a covert basis. Funding of ▓▓▓▓▓▓▓ (R&D study funds) was provided for the balance of FY-61.[20] General Greer chose the name GAMBIT for the new "black" program.

Raincoat, which dealt primarily with public information disclosure, did not completely resolve Charyk's desire to make GAMBIT covert. An anticipated potential weakness lay in the security aspects of normal Air Force contracting and financial reporting methods. It was military doctrine that the details of any activity involving the expenditures of public funds should be placed in public view, to assure practices of good stewardship. To preclude the security problems inherent in widespread financial disclosure—particularly to those who had no conceivable need-to-know—General Greer sought, and was granted on 5 January 1961, a contracting warrant directly from Secretary of the Air Force Dudley C. Sharp. With this warrant, Greer acquired authority equal to that reserved to the USAF deputy chief of staff for materiel; under such authority he could, where necessary, deviate from Armed Forces Procurement Regulations. Although Greer's procurement authorization was not particularly inhibited, Secretary Sharp advised him that "normal policies, practices, and procedures applicable to the Department of the Air Force" would be followed wherever possible. Greer was authorized to appoint contracting officers, to assign procurement authority to those officers, to approve time-and-materials contracts, to approve contractor overtime, to control government-owned industrial property, and to appoint and control property administrators. Such delegation of procurement authority to a program manager was unique.

During the first week of November 1960, a second set of key GAMBIT decisions emerged from the Pentagon. Charyk and Greer reconfirmed their determination to conduct GAMBIT as a covert reconnaissance operation and proposed to use the E-6 program as a cover for development of the system. Charyk agreed with Greer's suggestion that Kodak should develop the 77-inch camera under Project GAMBIT while General Electric would develop an orbital-control vehicle (OCV) and a suitable ballistic reentry vehicle for film recovery. By keeping the physical and environmental limitations of E-6 and GAMBIT compatible, it seemed possible to develop and test GAMBIT without any outward indication that such a program existed. The institution of rigid security controls over the entire Samos operation would greatly enhance the possibility of hiding the scope of the total program.[21]

About the same time, there was an effort by Air Staff elements, together with the Air Materiel Command and the Strategic Air Command, to continue to plan for normal military operation of Samos, of which GAMBIT was then considered to be a part.[22] To forestall such a move and to communicate clearly the planned objectives and operating principles of the Samos program, Charyk sent the Air Force Chief of Staff two memoranda of clarification. The first,[23] which was Secret, said that Samos "should be regarded as an R&D program aimed at the exploitation of various promising reconnaissance techniques" but that, until the completion of R&D, the nature of the system could not be determined and that "effective operational planning cannot be accomplished at this time." This action removed Samos from normal program documentation requirements, from monitoring by the AF Weapons Board, and from analyses by "the various panels, boards, and committees and directed that the intercommand Samos Working Group be dissolved."

In a concurrent but separate GAMBIT classified letter to Gen. Thomas D. White, Air Force Chief of Staff, Charyk identified a new philosophy for Samos.[24] He said it was essential to "maximize the reconnaissance take at the earliest possible date and to attempt to obtain such information in as low key a fashion as possible." He felt that the greatest chance of success would require establishing a "combination research, development, and operational program conducted under cover of research and development" and that there were compelling national policy reasons for avoiding any association with a military operational command "such as SAC."

The headquarters staff of the Strategic Air Command was understandably disturbed by this pronouncement. As operators of the free world's primary deterrent to foreign aggression, the staff assumed it should and would have a dominant share in acquiring strategic intelligence, whether by aircraft or satellite. For some years, SAC had worked hand-in-glove with the ARDC, the AFBMD, and the Space Systems Division to prepare for major operational responsibility in space reconnaissance. Gen. Thomas S. Power, now SAC's commander-in-chief, had watched the birth of Samos, during his ARDC days, and had cooperated in its growth. He had full expectation that a SAC team would launch Samos at SAC's Vandenberg AFB; that another SAC team would control the "bird" at a SAC Satellite Control Facility, and that a third SAC group would receive the intelligence product at a readout station. Power objected strongly (there was even a four-page telegram to the White House[25]) to "losing" Samos and urged his former protege, General Schriever (now commander, Air Force Systems Command) to join him in calling for the dictum to be revoked. Since Charyk's order had also cut AFSC out of the Samos pattern, unified Air Force opposition to the concept of an NRO developed immediately at the organization's inception.

Before Charyk's plan could become effective, program managers had to dispose of widely-dispersed evidence that a 77-inch camera development existed. The proposed Sunset Strip development program was so well-known that it would be necessary to invent and circulate a plausible motive for cancelling an essentially reasonable approach to satellite reconnaissance. Project personnel achieved this end by having SSD terminate the Kodak study contract for Sunset Strip, with the excuse that "review of recent proposals for E-6 camera reveals that future study in this area (77-inch camera) is not required."[26] Simultaneously, the Samos office drew up the first of its "black" contracts, authorizing Kodak to continue the development as a covert effort. Presidential reserve funds ("black" or "classified" funds) in the amount of ▓▓▓▓▓▓ were tentatively identified as the FY-61 program requirement.

The process of shifting GAMBIT camera development into secure facilities resembled that used three years earlier in sequestering CORONA work at Lockheed. As the Sunset Strip activity closed and personnel were nominally shifted to other Kodak projects, they actually moved into a new facility in a different building; there they were briefed on the fact that the project was very much alive, and resumed their work. Much the same procedure was followed with General Electric, although the fact that the E-6 and GAMBIT orbital control and reentry systems were closely akin, at least at first, greatly simplified the security problem.

By the morning of 7 November 1960, General Greer had briefed key officials of Aerospace Corporation (the systems-engineering support contractor for SSD), General Electric, and Eastman Kodak on the GAMBIT program, its objectives, and its relationship to E-6. He emphasized that the three principal contractors, plus the project office, would constitute a task force with the objective of developing and testing the GAMBIT system in the shortest possible time. There would be a good security shield: Lockheed, which ultimately became involved in the initial GAMBIT effort by virtue of the decision to use Agena as a stage in the launching system, supplied an essentially semi-standard vehicle; General Electric's cover would be the development of an alternate reentry body for the E-6; Kodak would rely chiefly on a "proprietary development" explanation; and Aerospace Corporation would operate under rigid "need-to-know" ground rules.

In December, thinking through the implications of several policy papers that had emerged since the National Security Council decision of August 1960, General Greer concluded that his real job was to "get pictures . . . in such a manner as not to precipitate a U-2 crisis in which the US might be constrained to discontinue Samos, and to insure the availability of systems which could covertly obtain needed photographs should even 'low key' reconnaissance operations become impossible."[27] His immediate task, he felt, was to create a real ability to operate a covert program, and his chief difficulty of the moment was that "the military system(s) for contracting and for disbursing money are very cleverly designed to frustrate a covert program."[28]

The elements of general policy under which SAFSP was to operate had been defined in February and appropriately circulated by the end of May 1961. On 29 May, a classified Headquarters USAF Office Instruction formally restated, for the benefit of the Air Force at large, the program rationale that had been adopted. For practical purposes, it was a formalization of Under Secretary Charyk's December 1960 memorandum to General White, neither expanding nor enlarging the instructions there defined. Considerably more important was a 3 April "Satellite Reconnaissance Plan" which defined in detail and in formal fashion the actual "policies, procedures, and actions to be applied . . . in order to achieve the . . . objectives of the national satellite reconnaissance program."[29] Those objectives were to enhance and protect the probability of "adequate and timely data collection" and to create a lasting ability to acquire reconnaissance information "in the event that circumstances should force limitations, reduction, or even elimination of overt flights."

The situation that prompted the covert effort was essentially that the overt objective of creating a US satellite reconnaissance system had been widely publicized, that regular flights ("overt and acknowledged") with military objectives were scheduled to begin in the near future, and that any indication of program success might provoke both political counteraction and a military response from the Soviet Union. The plan specified that:

> As a firm basic policy, there will be no "operational" overt satellite reconnaissance or any association of the program with an operational command for an indefinite time, and the overt satellite reconnaissance program will be brought to a fully operational status under cover of research and development, and operated indefinitely under this cover. The policy expressed in the 6 December 1960 Top Secret memorandum from the Under Secretary of the Air Force to the Chief of Staff, titled "Basic Policy Concerning Samos," will continue for the indefinite future.

Reflecting the urgency of technical efforts—in light of the political environment—the policy document contained a forthright statement of the need for more intensive control of project security and for the maintenance of "a viable covert effort which has the feasible capability of being sustained indefinitely after cancellation of the overt effort."[30]

Significantly, the objective of tightened security was to eliminate virtually all public references to military space programs and specifically to prohibit public disclosure of the flight test objectives or results of satellite reconnaissance. Within such an environment it seemed possible to culture a covert effort ". . . sustainable indefinitely in the wake of a forced public cancellation of the overt reconnaissance program, and which can meet all principal intelligence objectives of the overt program."[31] To that end, it was necessary to conduct the satellite reconnaissance aspect of the total Air Force military space program so unobtrusively that no indicators of the status of the overt program would surface in public. The covert program, of course, would be still more obscure—hidden even from those persons nominally cognizant of the extent and progress of the overt, but classified, effort.

While cover was generally needed in all parts of the GAMBIT program, its use to preclude disclosure was vital in contractor's plants—particularly in those performing unclassified, publicly disclosed, commercial work. The presence of the E-6 program effort at GE and Kodak did give local managers a means for "explaining" the presence of work in the plant, the movement of people, and the appearance of certain visitors. In detail, however, the cover did not always "work." At Kodak, for instance, the availability of facilities caused the E-6 program to be at one location—the Lincoln Plant—and GAMBIT in another, ███████████████████████████████████████ A significant number of GAMBIT people were in a guarded closed facility; their activity was explained locally as a company proprietary effort. At GE, E-6 work was also at a different location than GAMBIT work.

GAMBIT Varietals: Program 307, Exemplar, Cue Ball

By the spring of 1961, the E-6 and GAMBIT configurations were sufficiently different, both internally and externally, that the E-6 cover was wearing thin. At the same time, there was a growing probability (if CORONA continued to improve its capability) that E-6 would ultimately be cancelled. Greer had been concerned over the E-6 cover since its inception and now the problem was no longer academic. As early as December 1960, he had considered totally dissociating GAMBIT from the Samos effort; for a variety of reasons, it did not seem workable to hide GAMBIT as a "scientific satellite." But there was no easy or obvious solution. Finally, Greer—who had earlier initiated effective covert contracting on the basis that "everyone" knew the Air Force could not make significant purchases outside its insecure and involved review and approval channels—came up with the concept of a "null" program. A null program, in his definition, was one with no known origin and no published goal. Thus, a program with a highly classified and unidentified payload could purchase many parts of the system (boosters, upper stages, non-unique ground-support equipment, and many services) through normal channels. Viewed another way, if such "normal" items were procured through covert

~~SECRET~~

Handle via
BYEMAN-TALENT-KEYHOLE
Control Systems Jointly
BYE 140002-90

means it would dilute the cover and increase the possibility of disclosure. To further obfuscate the unwitting, Greer, who had been identified with Samos and, therefore, satellite reconnaissance, decided to show the "null program" as a responsibility of the Space Systems Division; this would indicate that the program was something other than reconnaissance. Such misidentification was easy, since SSD was, at that time, sponsoring a wide variety of programs, such as bombs-on-orbit, satellite interceptors, and communications satellites.

To provide "null program" support for GAMBIT, a "Program 307" was established in SSD in July 1961. On direction from the Air Staff, via the Air Force Systems Command, four "NASA-type" Agena-Bs were ordered for launchings scheduled to commence in January 1963. Subsequently, six Atlas boosters, configured to accept the Agena-B, were also ordered and purchased. In neither case was the hardware overtly assigned to a particular space project. To tie all this together, General Schriever, commander of AFSC, was directed by the Air Force vice chief of staff, in September 1961, to establish "Project Exemplar" (the name was classified "Confidential"), the purpose of which would be to provide four launchings from the Pacific Missile Range, beginning in February 1963.

The Atlases and Agenas ordered under Program 307 were assigned to Exemplar. To further "normalize" this overt effort and support cover, the GAMBIT Program Office stated requirements for the usual documentation; to do otherwise would have attracted unwanted attention.

The unclassified codename for Exemplar was "Cue Ball"; Air Force system No. 483A was subsequently assigned in December 1961. Not by coincidence, the program director was Col. Quentin A. "Q" Riepe who had previously been Midas program director (Midas was the infrared-detecting part of Samos). Riepe gradually assumed responsibility for GAMBIT from Col. Paul J. Heran who, as E-6 program director, had initially carried responsibility for GAMBIT. The transition was completed by February 1962.

Although it carried an Air Force priority of 1-A and a precedence of 1-1, Cue Ball was organized along the lines of a conventional SSD program (even though such "normal" channels and reporting lines were for cover purposes only); actual relations with higher authority would pass covertly through Greer's SAFSP office. It was particularly important, as Greer emphasized frequently in the early stages of setting up Cue Ball, that personnel prominently associated with the reconnaissance effort not be seen with Cue Ball personnel and that the Cue Ball people avoid any contaminating association with satellite reconnaissance. Not all Cue Ball assignees were cognizant of GAMBIT, so internal office security was an additional problem.

Misdirection continued successfully with Charyk's approval of the Cue Ball development plan and his formal authorization of initial funding at a level of ████. Key individuals in various offices in Headquarters USAF and AFSC had been alerted to the scheme and were presumably prepared to see that various budget, priority, and precedence authentications emerged promptly and satisfactorily. Initially, all went well. But, in a few weeks after Charyk's directive appeared, some of the carefully-laid cover began to flake away. Such unrelated events as attempts by non-briefed personnel to "straighten out" what appeared to them as anomalies; the problem of transferring funds from one part of the budget to another; and objections by non-briefed personnel to fully funding "objective-less" programs (at a time when apparently more deserving programs were underfunded) all caused problems and confusion and focused attention where none was wanted.

To resolve the matter, Charyk directed all GAMBIT funds in SAFSP, including Cue Ball, be carried under budget line item 698AL; thus, the program retained its high priority and precedence. There was some concern that the line item might be traced to Greer and identified as satellite reconnaissance, but this did not come to pass—another proof of Greer's original premise that nobody would suspect the existence of a "null program." As Robert Perry has stated: "Those in the inside of GAMBIT tended to seek complete normality as an avenue to inconspicuousness without appreciating that the regular Air Force establishment had been conditioned to accept uncritically any decision handed down, no matter how irrational. Rationality was not inherent in development decisions, nor logic a necessary ingredient of programming." He adds, "It was true that GAMBIT inhabited a covert atmosphere, and the procurement techniques and manufacturing practices invented for covert programs continued to be used, but in reality GAMBIT was a highly classified program without a publicly specified payload."[32]

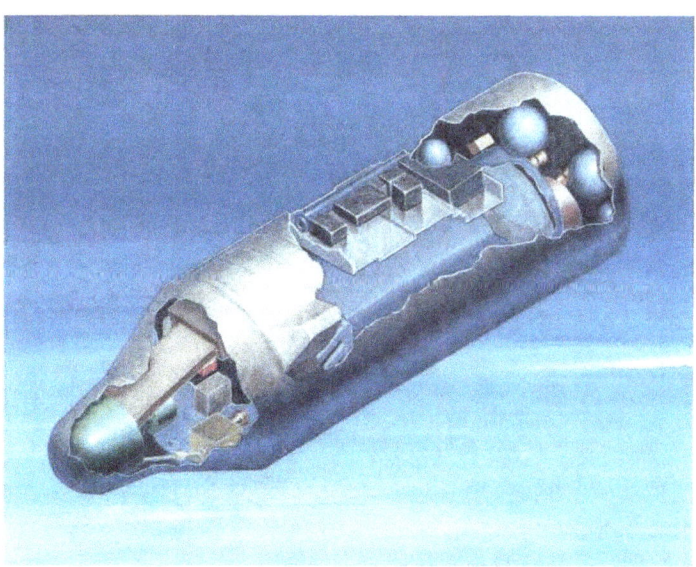

GAMBIT Orbital-Control Vehicle (OCV)—1961
(Artist's Concept)

GAMBIT System Characteristics

While these management and security evolutions were taking place, the GAMBIT development program was proceeding reasonably well. By January 1961, Kodak was under contract for the camera system. Similarly, General Electric's Space Division was under contract for both the orbital-control vehicle (OCV) and the recovery vehicle (RV). By mid-1961, GAMBIT had evolved into an approximately 15-foot-long, five-foot-diameter space vehicle.

The GAMBIT payload embodied a Maksutov f/4.0 lens (both reflecting and refracting elements) similar to an astronomical telescope with a 77.0 (±0.5)-inch focal length with a clear aperture of 19.5 inches. Its half-field angle, however, was much larger than that of an astronomical telescope, being 3.20°. This lens, when flown at a nominal 90-mile altitude, was to produce a ground resolution, at nadir, of from 2 to 3 feet. ▮▮▮▮▮▮

▮▮▮▮▮▮ GAMBIT was to carry 3,000 feet of 9.5-inch-wide, thin-base film through a strip camera, which would provide image-motion compensation by moving the film across an exposure slit at the same velocity that the projected image moved over the earth. When looking vertically, the camera would image a strip on the earth 10.6 nm wide. The system was capable of oblique pointing (accomplished by rolling the OCV) of ▮▮▮, could take either 15° or 30° included-angle stereo photographs, and was able to compensate for image motion over a slant range of 70 to 127 nm. It could be expected to take 300 to 600 stereo pairs, or twice as many monoscopic images. The planned weight of the total photographic system was 1,073 pounds.

The OCV was to be capable of varying the roll attitude from 0° to +▮▮▮ (with nominal roll-slewing rates of 0.25°, 1.50°, and 3.00° per second) and of performing 350 roll maneuvers at an average rate of one per second and an average amplitude of 30°. To perform pitch, roll, and yaw maneuvers, a freon cold-gas system with a total impulse of 8,000 lb/sec was used. A passive system of environmental control, with minimum use of heaters, was employed to maintain the lens bay between 65°F and 75°F and the stereo-mirror bay between 65°F and 78°F. Across the mirror face there was a 2°F gradient design goal ▮▮▮▮▮▮▮▮▮▮▮▮ normal to the mirror. Power for the nominal five-day flight was provided by storage batteries in the orbital-control vehicle.

▮▮▮▮▮▮▮▮▮▮▮▮▮▮▮▮▮▮▮▮▮▮▮▮▮▮▮▮ The command system received, accepted or rejected, and executed commands, both real-time or stored. Vehicle health data were to be telemetered by two VHF transmitters: one for real-time data, the other for recorded data. The transmitters could be switched by ground command, which provided a degree of redundancy.

Because of the relatively low altitude of the orbit (originally set for a nominal 95 nm for a five-day life but often flown below that orbital altitude) the GAMBIT spacecraft used two, ablatively-cooled, 50-pound-thrust engines (employing hypergolic propellants) to provide orbital adjust. These delivered a spacecraft velocity increment (Δv) of 400 feet per second and had a total impulse of 60,000 pound seconds. The reentry vehicles were ultimately very similar to the proven CORONA configuration, although, during the early days of the program, a new and somewhat larger configuration of a different shape and ballistic coefficient was carried to the early development stage.

The initial GAMBIT launching vehicle was an Atlas Agena-D. The Atlas stage, a direct derivative of the "stage-and-a-half" ballistic missile, used a total of 123 tons of liquid oxygen and RP-1 fuel (a refined kerosene) to power two booster engines—each generating 154,500 pounds of thrust—and a 57,000-pound-thrust sustainer engine. The Agena-D upper stage (which became the Standard Agena) used 13,234 pounds of hypergolic propellants (unsymmetrical dimethyl hydrazine and inhibited red-fuming nitric acid) to power its 16,000-pound-thrust engine.

After exposure, the camera's film was rolled up in the recovery vehicle (RV). At the end of the mission, the RV was separated from the OCV, spun up on its axis of symmetry by a cold-gas system, and then given appropriate retro-velocity to deboost the RV. Initial parachute deployment occurred at 55,000 feet, followed by separation of the ablative shield. The final step was aerial recovery.

Numerical Summary of GAMBIT-1 Payload[33]

Photographic Output Data

Ground resolution (vertical photography)	2 to 3 feet
Lens-film resolution	
Scale of photography	(95 nm altitude) 1:90,000
	(70 nm altitude) 1:66,000
Width of photographed strip	
(vertical photography, 95 nm alt)	10.6 nm.
(vertical photography, 70 nm alt)	7.8 nm.
Scene width on payload film	8.518 inches
Scene length on payload film	Variable
Scene length on ground	Variable
Number of photographs	300–600 stereo pairs or equivalent amount of continuous strip photography

Payload Package

Weight	
Camera Payload components in OCV	1,079.5 lbs (w/o film)
Camera Payload components in SRV	22.6 lbs (w/o film)
Film	52.0 lbs (3,000 feet)
Total	1,154.1 lbs
Dimensions of Camera Payload	
Maximum diameter	54 inches
Length	190 inches (front of recovery cassette to aft mounting plane)

Payload Camera

Camera Type	Strip
Exposure (Nominal)	1/400 sec with 0.0085-inch slit
	1/200 sec with 0.0169-inch slit
	1/100 sec with 0.0338-inch slit
Number of slits	3 photographic; 1 orbital test; 1 ground test

Numerical Summary of GAMBIT-1 Payload (continued)

Lens

Type	Maksutov
Focal length	77.0 (±0.1) inches
Aperture	19.50-inch diameter
Half-field angle	3.2°
Filter Type	Band-L Type 10
Spectral angle film and filter	500–700 millimicrons
Focus Adjust Type	Single grid, single detector, and channel with rotating focus filter. Range +0.010 inch
Focus drive	Platen drive by d.c. motor at 0.00025 inch/sec (nominal)

Payload Film

Type	Kodak High-Definition Aerial Film (Estar Thin Base) Type 3404
Dimensions (±0.010)	
Width	9.460 (±0.005) inches
Length	3,000 feet
Thickness	0.0030 (±0.0003) inches
Base type	Polyester
Weight	52 (±3) lbs (3,000 feet)
Roll dimensions	
Core diameter	4.25 inches
Outer diameter	13.00 inches
Film tension	3.00 (±0.25) lbs

Image-Motion Compensation (IMC)

Film Drive Velocity Range (70 nm alt)	[redacted]
IMC Tolerances	
Average Velocity	[redacted]
Smoothness (RMS of velocity transients)	
Velocity Transient (maximum)	
Number of speed steps	64 + OFF/ON
Speed change per step	1% of previous step
IMC Design Parameters	
Obliquity Range	
Stereo Aim Angle	-15°, 0°, +15°
Altitude Ranges (all obliquity, 0° stereo)	
Camera Payloads 1–6	95 (±12) nm
Camera Payloads 7–10	72 (±9) nm
Camera Payloads 11–54	81 (±11) nm

Optical Aiming

Stereo Positions	-15°, 0°, 15°
	Crab
Positions	0° to +3.5° or 0° to -3.5°
Steps	8
Angular Interval	0.5°

Obliquity Aiming (by OCV)

Position Range	[redacted]
Angular Interval	
Angular Rate	
Roll settling time	

GAMBIT/Atlas Agena-D Launching

Land Recovery Versus Aerial Recovery

At the inception of the GAMBIT program, land recovery had been directed; it had appeared to those at high levels of government to be more straightforward and less of a security risk than the ocean recovery used by CORONA. But several factors caused the GAMBIT recovery capsule to grow heavier and more complex. First, it had to withstand parachute failure and not rupture on land impact. Second, it had to be locatable at considerable in-track and cross-track error distances over rugged terrain and in bad weather, so it needed a more sophisticated and sturdy beacon than envisaged originally. Finally, even though the population density of the chosen recovery site (Wendover AFB, Utah) was low, the need to be confident of avoiding even a few populated areas demanded better accuracy and certainty of performance than had been anticipated initially.

Although early on, General Greer had favored land recovery, he soon came to the view that the essentiality and practicability of land recovery had been over-emphasized. As he watched the improving capability of the CORONA RV and the good performance of the overwater recovery system, the value of land recovery diminished in his mind. On several occasions during the first year-and-a-half of GAMBIT's development, he informally discussed with Charyk the land-versus-overwater issue and the possible use of the CORONA RV on GAMBIT. By July 1962, the GAMBIT RV had grown about 500 pounds over its specified weight. While the Aerospace Corporation had earlier studied aerial recovery for the GAMBIT capsule, the method had been considered impractical, since the capsule's forecast weight, even then, exceeded the capability of the C-119 recovery aircraft.[34] Greer believed that the basic reason for distrusting aerial recovery—fear of loss or compromise of the capsule—had abated substantially since the initial program directive of 1960; the overwater aerial recovery capabilities developed for CORONA now contained provision for dealing with nearly all foreseeable contingencies. At the same time, the possibility had not diminished that a land-recoverable capsule (particularly a new and unproven model) might fall in either Mexico or Canada, or hit a populated area in the United States.

In July 1962, Greer again raised these issues with Charyk, who agreed that alternatives to land recovery should be studied. At the time, Greer was concerned with GE's progress on the GAMBIT RV. He decided that the feasibility of using a CORONA-like RV on GAMBIT should be studied and he personally directed Hilliard Page, GE's general manager, to do so. He also ordered his program director, Colonel Riepe, to study the matter. If the overwater alternative should prove feasible for GAMBIT, its recovery problems would be resolved. Preliminary results from GE were encouraging and Greer reported to Martin and Charyk that use of a slightly modified CORONA capsule would provide "a vastly simpler scheme for recovering recorded data for certain special projects."[35] Riepe and members of the GAMBIT Program Office were less enthusiastic and found it difficult to fault the current approach; moreover, although Riepe himself was cognizant of CORONA, his people were not, and even Riepe did not know CORONA's details. Offsetting his natural reluctance was the attraction of potential savings on weight, cost, and schedule.

SECRET
Handle via
BYEMAN-TALENT-KEYHOLE
Control Systems Jointly
BYE 140002-90

Finally, the situation began to change. Based upon a briefing by Riepe to Charyk on 24 August 1962, a decision was made to conduct the first system tests over the Pacific Recovery Area, using aerial retrieval as the primary recovery method. In addition, "all development activities on the present land-impact vehicle will be reduced to the minimum expenditure rate," and alternatives would be prepared for Charyk's further study.[36] On 18 September 1962, it was decided to terminate the land-recovery program and to change to CORONA's H-30 recovery vehicle configuration.

In retrospect, this partly intuitive action by General Greer was a key—and possibly essential—ingredient to the success of the GAMBIT program. The overweight land-recovery RV could have jeopardized the entire program; further, the probability of timely and within-budget development of a completely new recovery capsule during the early 1960s was not high. To underscore this, a major factor in the cancellation of the Samos E-6 program—which occurred on 31 January 1963—was the continued operational failure of its GE-developed recovery capsule.

Payload Development at Eastman Kodak

While these major decisions were being made, the payload and OCV developments were progressing. At Kodak the challenges were being met. The camera subsystem consisted basically of optics, film-handling, and supporting mechanisms and electronics. The optics were to be larger and lighter than any previously built for space use. The primary mirror and the stereo mirror were to be made by novel techniques. The so-called "blanks" (unground and unpolished mirrors) were made by the ▓▓▓▓▓▓▓▓▓▓▓▓▓▓ for Kodak. Using large boules of very pure fused (amorphous) silica glass, face and back plates were cut, as were the interior pieces, which were thin, notched, quasi-rectangular plates joined in an ▓▓▓▓▓▓ fashion. The mirrors were assembled with the back plate supporting the ▓▓▓▓▓▓ section, surrounded by side plates, with the to-be-finished face plate on top. This assembly was placed in a large furnace where it was heated just to the melting point of silica, at which point the various pieces fused to each other. The fusion operation was delicate: too long a time or too high a temperature would make the intended structure a partially molten blob, while too low a temperature or too short a time would prevent the parts from fusing sufficiently to provide structural integrity. After the fusion step, various tests were made to determine the percent of intended fusion that had actually taken place and to establish the geometry of any voids. Criteria for acceptance or rejection of the fusion process for the assembled blank had already been established.

After some early failures, these large, lightweight blanks were successfully manufactured by Corning and shipped to Kodak for figuring and polishing. To perform this work, Kodak had prepared a special facility in its ▓▓▓▓▓▓▓▓▓▓▓▓ where new, large, grinding and polishing machines had been built. Well-proven techniques were used and success was largely a question of scale, as

Detail of ▓▓▓▓▓▓ Mirror Construction

well as proper concern for the fact that the structure being ground and polished was more delicate than the usual piece of solid glass. An integral part of the figuring and polishing step was the need for repeated testing to insure achievement of the desired optical figure. The optical-figure-error budget required that the spherical primary and flat stereo mirrors be accurate to a root-mean-square value of one-thirtieth of the wavelength of light ($\lambda/30$) as well as a peak-to-peak value of the same magnitude. At that time—the early 1960s—laser light sources were first becoming available, offering optical test engineers something new and useful—a coherent light source. During the early phases of GAMBIT, optical elements were tested by white-light knife-edge techniques; later, laser interferogram methods were used. To test the entire assembled optics, full-aperture auto-collimation was employed. While optics development and testing were not without problems, most of those were quickly resolved. Similarly, the film-handling hardware, which used a looper system conceived during Kodak's early work on a camera for the A-12 OXCART aircraft, posed no unique problems. The principal mechanisms used by Kodak in the GAMBIT payload were either structural or displacement (motor devices). The principal structures were the mirror mounts and the optical barrel. The latter was essentially a ▓▓ The mirror mounts were unique in that they had to hold both primary and stereo mirrors gently—so as not to introduce distortion—but firmly—so as to withstand shipping and launching loads without any displacement. The support electronics (which interfaced with the command programmer in the OCV) were used to control film-drive velocity over the exposure slit, to control stereo-mirror movement and placement, and to position the desired exposure slit. To provide the proper thermal environment for precluding distortion of the optics, the payload had its own thermal-control subsystem, which interfaced with that of the OCV. The concept, which worked well when the OCV met the interface requirements, was to open or close the payload bay door to keep the payload just a little cooler than its desired ambient condition and then, by adding heat to the optical barrel and mirror back plates with strip heaters, bring them to the desired temperature and temperature gradient.

The program at Kodak was directed by Dr. Frank Hicks, whose principal team-members were James Mahar, systems engineering; Leslie Mitchell, payload design; John Sewell, test and support equipment; and Don Stevens, support and administration. The entire project was located in Kodak's ▓▓▓▓ and reported to the director of Special Projects—originally, Dr. Kenneth MacLeish, who was replaced, in 1961, by Dr. Frederic C.E. Oder. Earlier, when Oder was in the Air Force, he was the original WS-117L project officer and was witting of the entire CORONA effort. The Special Projects organization reported to Arthur Simmons, director of research and engineering of the Apparatus and Optical Division. Because of its national importance, the GAMBIT project was given a high priority on acquiring people—not only within the A&O Division but company-wide. A special organization was set up to handle the program's physical and personnel security needs.

With the decision to use the CORONA RV in GAMBIT, it became apparent to Oder that key people in his GAMBIT program organization had the need to know certain aspects of CORONA, so they could make the use of that system's technology. He arranged for Lt. Col. John Pietz, of Greer's security office, to provide authorizations and CORONA briefings to a few GAMBIT project people at Rochester. This allowed Kodak to make much better use of the RV than was otherwise possible. For example, based on earlier Samos experience, Kodak had originally planned to keep the film path pressurized, including the film chute and take-up cassette; it now learned, from CORONA information, that an unpressurized film path could be used. The effect of this information was to simplify the take-up cassette and allow the GAMBIT film load to be accommodated within the CORONA RV.

Orbital-Control Vehicle Development at General Electric

The OCV development by General Electric, in its Valley Forge, Pennsylvania, facility, was not an easy assignment. The effect of failures in such varied components as harnesses, power supplies, batteries, command systems, horizon sensors, rate gyros, environmental doors, and pyro devices, multiplied the tasks originally envisioned for the OCV, jeopardizing attainment of the original cost and schedule goals. GE decided that it could most likely locate potential failures in flight hardware by doing comprehensive thermal-vacuum and vibration testing at a complete OCV level-of-assembly, and routinely checked out all vehicle components at that stage. While box-level testing may have been done, the official GE 206 Program Report made no mention of a comprehensive test program of that type. OCV problems were not unexpected, as seen from this quotation from a GE 206 Program Report:

Some indication of the scope of the task can be seen from the fact that the vehicle, less camera payload, contained 206 components and 80 black boxes, including 4,000 mechanical piece parts, 39,000 electronic piece parts, 575 harness connectors, and 5,000 harness wires. These totaled approximately 160,000 potential defect sources that had to be screened out in every vehicle, often in the face of tight launch schedules.[37]

Later industrial experience showed that it was less costly, in terms of labor (dollars) and schedule, to test rigorously at the lowest level-of-assembly, rather than to locate failures when the complete flight vehicle had been assembled.

The nature of the OCV problem was to be dramatically characterized later in a 29 August 1967 letter from Brig. Gen. John Martin, SAFSP, to Dr. Alexander Flax, DNRO, summarizing the initial GAMBIT program: "[W]ith the exception of one Agena failure and one Atlas failure . . . all of the mission catastrophic failures and most of the other serious failures were in GE equipment."[38]

As stated earlier, many of the problems that arose with the OCV during its development phase were found during vehicle-level tests and were resolved by redesign and rework. An example of a component requiring extra work was the horizon sensor, which sensed the earth-sky boundary, which was essential for the vehicle to achieve proper orientation. When GE's horizon-scanner development got into trouble in 1962, the GAMBIT Program Office started two other efforts toward a solution: one at Eastman Kodak, the other at ▓▓▓▓▓▓▓▓▓▓▓▓ All three were carried until September 1962, when SAFSP, receiving results from the three approaches, decided in favor of ▓▓▓▓▓ Both the GE and Kodak developments had, by then, proved their utility; however, neither was better than the ▓▓▓▓▓ instrument and both were more expensive. Further, the ▓▓▓▓▓ sensor operated over the widest target-temperature range, which made it more effective in a winter environment. On 17 September 1964, GE was advised contractually that ▓▓▓▓▓ sensors would be government-furnished for GAMBIT.

Not to be outdone, about this time EK also had problems in two areas. One involved the means by which the large silica mirrors were attached (cemented) to their metal cases; the other resulted from using incorrect fluid in a dash-pot in the platen-drive, causing the film to move irregularly over the exposure slit. These problems were resolved but they did add to existing pressure on hardware delivery and flight schedules.

Prompted in part by hard questioning during an October 1962 meeting with the President's Foreign Intelligence Advisory Board (PFIAB) and the Special Group of the National Security Council, Charyk characterized GAMBIT as "imperative" and urged that it be developed with a "maximum sense of urgency," noting that the "extreme political sensitivity of any other method of obtaining such photography"—to wit, overflights by U-2 or OXCART aircraft—made it essential that "no reasonable steps should be omitted to guarantee GAMBIT's success at the earliest possible time." Discouraged by the rate of GAMBIT progress, Charyk requested an exhaustive review to locate any problems remaining in the program. He emphasized that resolution better than the two-foot requirement of 1960 was desirable. He also cautioned that money was not unlimited and that greater management talent, rather than more funds, should be applied to the program.[39]

In all probability, the prevalence of cost over-runs (particularly at General Electric), the threat of new schedule slippages, and the increasing cost of the GAMBIT program prompted Charyk's concern over the future of the development. He was disturbed by the possibility of additional schedule slippages, since only GAMBIT offered hope for discovering whether the Soviets were actively preparing military forces for use. The coincidence of Charyk's anxiety with the start of the Cuban missile crisis of 1962 should also be noted; even though the United States did not have clear evidence that Soviet nuclear missiles were being installed in Cuba until the second week of October, concern over that possibility had been mounting since August.

General Greer was fully aware of circumstances that had moved Charyk to his exhortation and possessed no convincing evidence on which to base a rebuttal. The Samos E-6 program was in grave technical trouble in October 1962, having experienced four recovery-vehicle failures in as many flight attempts. Because of a succession of misfortunes, it had been necessary to cancel each of the major photo-reconnaissance programs assigned to SAFSP in the original Samos program, except for E-6[40] and GAMBIT. True, the most obvious defect in GAMBIT design had been eliminated with the decision to adopt aerial-recovery techniques and to use the CORONA recovery vehicle. But the prospect of continued GAMBIT slippage was still very real and there was no strong confidence that the complex camera system would function properly during its early flight trials.

GAMBIT's Final Home—SAFSP

On 5 October 1962, Greer, with some reluctance, told Charyk that the most certain way to strengthen GAMBIT management would be to transfer custody of the program from the Space Systems Division to SAFSP. The desirability of this transfer had been examined in detail as early as July 1962. By October, Col. J. W. Ruebel, Greer's special assistant, had worked out the basic details of a transfer plan and had composed a rationale for public consumption. Greer expressed a desire to keep Colonel Riepe in charge of the program.

Greer told Charyk that moving the program into SAFSP would give GAMBIT the prestige of the Office of the Secretary of the Air Force, although it did seem possible that identification of GAMBIT with reconnaissance objectives might follow. In Greer's eyes, that possibility was not a disqualifying handicap. He reminded the under secretary that the United States had constantly maintained the basic legality, under international law, of satellite reconnaissance and that the nation had never denied the existence or employment of orbiting camera systems. The chief purpose of concealment now, he suggested, was to cloak the scope and success of such operations. That much could be done within SAFSP. In the remote possibility that national policy shifted, it would be difficult to continue any effort even indirectly associated with reconnaissance objectives.

Greer was not optimistic about the prospect of improving the quality of GAMBIT photography, at least in the first several flights. He told Charyk that the original resolution requirement—two to three feet—would very probably be satisfied, although he knew that not all experts agreed with him on that score. Greer cautioned that results from the first few flights might not bear out his conviction that GAMBIT would indeed prove itself; past experience with new space vehicles (into which category the General Electric orbital-control vehicle fell) did not encourage strong optimism. As for priorities and emphasis, Greer noted that it was difficult to convince either contractors or military personnel involved in administration of this program that it enjoyed any special priority or importance, since the one infallible indicator of status—timely and adequate funding—had been consistently absent.

Early in November 1962, General Greer repeated his suggestion of transferring the entire 206 Program to SAFSP. Answering earlier objections, he explained to Charyk that such a move did not imply "surfacing" the development or acknowledging its reconnaissance objectives: the payload would remain covert and procurement would be "black." Moreover, the cover plan devised in SAFSP promised to perpetuate the legend that Program 206 (Cue Ball) was in some way related to a bombs-in-orbit program. The explanation for project transfer from SSD to SAFSP did not need to be either complex or particularly sophisticated; a straightforward statement that program priority placed it under the direct control of the Secretary of the Air Force would satisfy those who did not know that covert programs were being conducted within the Air Force. Greer reasoned that those aware of the existence of clandestine activities would deem it unthinkable to move a concealed reconnaissance program into a reconnaissance organization and would be more firmly convinced than ever that Program 206 had some mission other than satellite reconnaissance.

"Children or half-wits, if they care, will most likely reason directly to the correct deduction, i.e., if it's assigned to SAFSP, it's reconnaissance. Inasmuch as we will do nothing to confirm this, and we will insure that some actions are apparently inconsistent with the hypothesis, I think there is a good chance of fooling—or at least confusing—the professional espionage agent, who is presumably neither a child nor a half-wit."[41]

There was another consideration which influenced GAMBIT's transfer to SAFSP. The implementation of Raincoat (discussed earlier) by Department of Defense Directive 5200.13 had placed all military space programs in a "no publicity on payloads" and "special access, must-know" category. Individual access lists were being maintained for each program and program information was being confined to those having an approved need-to-know. Random numbers were substituted for previously-used popular names and launching announcements were restricted to a bare statement of the type of booster and the date of the operation. In such circumstances, it was no longer possible to identify a Samos payload solely from the fact of launching security; all military space launchings were being conducted under tight security provisions. Thus it was increasingly difficult for someone who did not have program security access to acquire information about most Air Force space programs; to a degree all cover stories were now somewhat redundant.

Greer's arguments were effective. By 20 November 1962, Charyk concurred in the "desirability" of transferring 206 to SAFSP. Maj. Gen. O.J. Ritland, who was now part of the Air Force Systems Command (AFSC) headquarters staff, was called in to brief Maj. Gen. Ben I. Funk, now commander of SSD, on the realities of the situation.

The loss of the 206 Program was resisted and resented by the Air Force Systems Command. General Schriever, now commander of AFSC, had been a major force in establishing the Air Force space program. To him, the fact that any Air Force space activity was not under AFSC's management was "not right," and, in 1962, under his leadership, AFSC made determined but unsuccessful efforts to regain "ownership" of all Air Force space programs. Ironically, the streamlined "Gillette" management concept that Schriever had enjoyed as commander of AFBMD, had lost its attractiveness to him, now that high-priority space programs were reporting directly to the Office of the Secretary of the Air Force.

While transferring Program 206 to SAFSP, Charyk also considered strengthening the program's leadership. He had been strongly impressed by Col. William G. King, Greer's technical planning officer, who was conducting forward-looking studies (including the VALLEY program—an early effort at developing a ▓▓▓▓▓ search system). He knew that King had been associated with satellite reconnaissance for nearly 10 years, beginning with early Wright Air Development Center days and culminating in assignment to the AFBMD as Samos Program Director (1959-60). Charyk knew King as a consistent pragmatist, who had been one of the first to call for a film-recovery Samos (accepting the reality of then-current read-out rate limitations). As this series of favorable impressions were recalled, Charyk proposed King to Greer as a clear best-choice to direct GAMBIT. Fortuitously, when General Funk heard of this possible change, he informed Greer that he would like to assign Riepe to a new SSD development program. King became the GAMBIT manager on 30 October 1962.

One of Colonel King's first actions, after assuming GAMBIT management, was to advise General Greer that the current design of the adapted recovery capsule represented much more of a change than Greer had intended. Greer had ordered "minimal changes only" to the CORONA capsule; now he emphatically endorsed Colonel King's recommendation that the original intent of the modification be reinstated and that the General Electric development effort be redirected accordingly. King carried out the order: meeting with key GE officials, two days later, he defined the objective of the capsule change. Cross-briefing GAMBIT people on CORONA—a continuation of the process earlier begun at Eastman Kodak—helped restrict GE's engineering approach to one of (only) limited modification of the recovery vehicle.

Colonel King had also made it clear to EK that system changes were to be minimal and that any changes in configuration of the CORONA capsule would need his personal approval. By all indications, King expected external changes

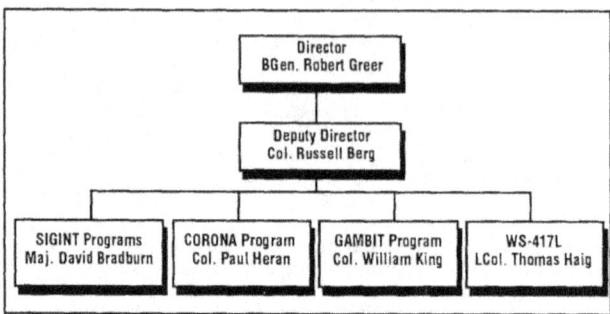

SAFSP Organization in 1963

to be slight. The general policy, he added, was to use flight-proven components wherever possible, keeping all changes to a minimum, but altering the details of payload configuration as essential to the requirement for limiting external change.

Hitch-Up, Roll-Joint, and Lifeboat

At this time, the GAMBIT Program Office was still concerned over the OCV's attitude-control subsystem. As a conservative measure for improving the probability of early fight success, it was planned that the Agena, for the first three flights, would remain connected to the OCV for most of its orbital life. In this "Hitch-Up" configuration, the Agena would provide attitude-control for the spacecraft. Later, the OCV would be separated, to demonstrate its own attitude-control capability. It was anticipated that the mature Agena was more likely to perform correctly than was the new OCV. It was important to General Greer and his team that early flights obtain good quality photographic imagery, even though flying in the Hitch-Up mode did not allow demonstration of full-system capability of the OCV (and permitted only near-nadir photography).

There was more to Hitch-Up than met the eye. An elaboration of the scheme involved use of a roll-joint coupling (invented for an interim high-resolution satellite known as Project LANYARD and its KH-6 camera)[42] between the stable spacecraft (Agena) and the camera. Should the orbital-control vehicle prove generally unreliable, it might be possible to introduce the LANYARD roll-joint between the Agena and the payload end of GAMBIT, eliminating reliance on the stability and control elements of General Electric's orbital-control vehicle.

On 29 November, General Greer presented the results of a preliminary analysis of the Hitch-Up and roll-joint ideas in a meeting with Charyk, who showed interest. Greer then drafted an authorization for continued study of these options and sent it to Washington for endorsement. On 30 November, the second major change (in two months) to GAMBIT was tentatively approved. Colonel King took the view that uncertainty over a successful demonstration of the OCV stabilization system mandated proceeding with development—at LMSC—of a roll-joint for GAMBIT and, in a January 1963 policy paper, authorized GE and LMSC to prepare for roll-joint use, should that become necessary. The plan called for LMSC to deliver three roll-joint modules before the end of the year.[43]

Bringing the roll-joint into the GAMBIT program raised a security problem. The roll-joint was quite unknown to most GAMBIT people and it seemed unwise to disclose the existence of the LANYARD effort to large numbers of GAMBIT workers. So Charyk's message of 30 November (actually written by General Greer) contained the "suggestion" that Greer contact Lockheed about the roll-joint because Charyk believed "a similar idea was once proposed and possibly designed in connection with another space program."[44] The kernel of a cover story was outlined: Lockheed would be empowered to "develop" the earlier "idea," delivering finished roll-joints to GAMBIT as though they were new items with no relationship to any other reconnaissance program.

As the result of a full-scale technical review held by King, a further measure for insuring the success of GAMBIT flights was proposed to Charyk on 14 December 1962: a recommendation to incorporate a backup vehicle stabilization system and limited command capability in GAMBIT. Named "Lifeboat," or "BUSS" (Back-up Stabilization System), this was another design approach originated in the CORONA program. It included a separate magnetometer, a separate cold-gas stabilization system (including gas supply and controls), and an independent reentry-command receiver and associated circuitry. All of these were completely independent of the main OCV subsystems and could be activated if the primary reentry systems failed. The magnetometer referenced the vehicle's spatial orientation to local lines of magnetic force; with this information, the vehicle could use Lifeboat's gas-stabilization capability to orient itself properly for separation and de-boost of the RV. The roll-joint provided added assurance of proper attitude control. With these changes, GAMBIT, in December 1962, was a much more realistic development than it had been four months earlier.

Finally, although he agreed that the most vital initial objective of GAMBIT was to return "one good picture" (Greer's frequently-stated goal), Charyk nevertheless insisted that all flights subsequent to the first should be programmed to return useful pictures of pre-selected intelligence targets. He specifically rejected the concept of a step-by-step approach to an operational configuration through research and development improvements. His philosophy was key to the reason for incorporating a roll-joint development: if it were necessary to rely on the roll-joint—because of failure of the GE orbital-control vehicle—the GE effort could be discontinued. Degradation of picture quality was a probable consequence, but the degree of degradation could not be accurately estimated. The OCV was being built to have more precision and greater granularity of roll-position than that available from the existing roll-joint. The LANYARD roll-joint system could provide only 100 stereo pairs of pictures of selected targets during a single mission—about one-third to one-fourth of the current expectation for the GE vehicle and one-sixth of the original requirement. The roll-joint was designed to permit shooting at angles as great as 30° from vertical, with intermediate settings every 5°. In late 1962,

NRO Director Brockway McMILLAN   USAF Col Quentin A. RIEPE

GAMBIT program people were concerned about compensating for smear and image-motion-compensation errors when the roll-joint was in use.

On 19 December 1962, Charyk formally authorized Lifeboat, Hitch-Up, and roll-joint additions to GAMBIT. Lifeboat was to be a permanent part of the total system, Hitch-up was to be incorporated in the first four vehicles (but a determination on use would be made on a flight-by-flight basis), while the roll-joint was to be developed "as a *bona fide* operational substitute for the OCV roll system." "Black" costs, all for the roll-joint, came to ▇▇▇▇. "white" costs, covering Lifeboat, Hitch-Up, and remaining roll-joint expenses, totaled ▇▇▇▇.[45]

In December 1962, Greer approved King's proposal to delete a portion of the elaborate test program that had been planned earlier. The first GAMBIT launching was still scheduled for July 1963. Greer realized that reducing the scope and number of development tests posed a risk but, by the same token, another cost overrun or a further schedule slip would also threaten the entire program. Offsetting the risk was the advantage of using proven hardware (Lifeboat and the CORONA RV), which provided greater assurance of recovery success and insurance against catastrophe. (Later, Hitch-Up results were to show that Greer and King had been too pessimistic about the stability capabilities of the Agena and, perhaps, too exacting in their requirements for camera stability.)

On 1 March 1963, Dr. Charyk resigned to become president of the newly-formed Communications Satellite (Comsat) Corporation. He was replaced as DNRO (and as under secretary of the Air Force) by Dr. Brockway McMillan, of Bell Telephone Laboratories.

Section 3

GAMBIT Operations: The Early Flight Program

GAMBIT flight vehicle No. 1⁴⁶ lifted from its Vandenberg Air Force Base launching pad on 12 July 1963, thrusting its way toward a position 110 nm deep in outer space. General Greer, Colonel King, and their associates knew it would be 90 minutes before they would have proof that the bird was in a proper polar orbit. After that positioning, they would feel limited initial assurance; then five orbits would be counted off before the payload began an inquisitive search of the USSR. After nine "working" passes, the satellite would be called back to earth, after ejecting the RV carrying a cargo of engineering data and overflight photography. The shortened flight plan was in deference to the fact that GAMBIT No. 1 was a new bird; at this stage, demonstration of flight skills was more important than performance as an intelligence-gatherer.

On the 18th orbit, a ground station commanded GAMBIT back toward earth and a C-119 aircraft, waiting near Oahu, swept the parachuting reentry capsule out of the sky.

The first GAMBIT "try" was a success; although only 198 feet of film was exposed, the average photographic resolution was 10 feet and some of the best was close to 3.5 feet. Greer's mandate to King had been "One good picture," with emphasis on "good"; the GAMBIT team had more than met that goal.

The second flight took place on 6 September 1963. Although purposely limited to two-plus days on orbit, it still delivered 1,930 feet of exposed film and covered ▓▓▓ the targets specified by USIB's Committee on Overhead Reconnaissance (COMOR). The best ground resolution—2.5 feet—met the basic design specification, greatly reassuring GAMBIT's optical engineers. This particular resolution value, translated into lay terms, meant that photointerpreters could distinguish such items as aircraft nacelles and small vehicles. For the first time, a satellite-reconnaissance system had produced pictures at resolutions previously obtained only by reconnaissance aircraft.

On the debit side, a member of the intelligence "user community" could grouse, with reason, that 1,930 feet of film and coverage ▓▓▓ COMOR-assigned targets hardly called for a festival. And engineers were distressed by the fact that the orbital-control vehicle had not functioned well enough to demonstrate pointing accuracy.

The third flight, launched on 25 October 1963, was disarmingly smooth in all respects. In general, the photo quality was excellent: this was the first satellite mission to identify people on the ground ▓▓▓ the scene being a football game. This flight was also the first to use color film. The return was generally degraded by improper exposure, and, like the previous GAMBIT returns, a very limited amount of true intelligence information was produced. Emphasis still lay on engineering validation of the satellite.

On the next flight, the Lifeboat back-up stabilization system was installed (its control gas and command circuitry independent of the orbital-control vehicle's primary stabilization system). GAMBIT No. 4 was launched on 18 December 1963. The goal of increased emphasis on actual intelligence-gathering was not realized; an unstable "rate gyro" triggered massive instability in the OCV and there was no point in even attempting photographic operations. Lifeboat did its best to correct the situation; although overwhelmed by the severity of the problem, it performed well enough to become a standard item on the next 32 flights.

GAMBIT-1 Flight Summary—1963

G-1 No.	Date	Targets Covered	Best Resolution	Days on Orbit	Remarks
1	12 Jul 63		3.5 ft.	1.1	
2	6 Sep 63		2.5	2.1	
3	25 Oct 63		3.0	2.1	
4	18 Dec 63		—	1.1	Loss of orbital-control gas

The fifth flight, on 25 February 1964, was as unproductive as the fourth. A series of strange command faults and errors exposed serious flaws in the OCV and in controller communication procedures:

- Telemetry indicated that the roll and pitch gyros had not uncaged. Actually they were functionioning very well, but
- The ground controllers, misled by the telemetry, sent a new "uncage" command. This forced a serious yaw problem; however,
- The yaw problem was corrected on revolution 18, but—
- A film "cut-and-load" signal had already been sent on revolution 16.

GAMBIT No. 6, launched on 11 March 1964, also had roll-joint malfunctions but redeemed itself with targets acquired (a new record, by far) and generally excellent photography. The Intelligence Community was still dissatisfied: its members had become accustomed, rather quickly, to the idea that resolutions of two to three feet were obtainable by GAMBIT and naturally wanted this best capability exercised against all targets. An interesting sidelight to the flight was that, after capsule ejection, a low-altitude experiment was conducted for seven revolutions at 70 miles, with no apparent spacecraft problems. It was hoped that some low-level passes could be made on subsequent flights to push best resolution beyond the excellent then-current values.

GAMBIT No. 7, launched on 23 April 1964, performed better than No. 6, particularly during its two days at low orbit.

The Hard Times of 1964

The eighth GAMBIT, launched on 19 May 1964, showed more control problems: attitude reference was lost mysteriously on the 15th revolution and restored on the 25th. Resultant photography was good in quality but disappointing in quantity. There were similar control problems with GAMBIT No. 9, launched on 6 July 1964, which returned no useful photography. It was finally noted, however, that control problems occurred only when the satellite was over the Antarctic; a possibility existed that the control's sensor could not distinguish between the temperatures of Antarctica and outer space, during winter months. Since redesign of the sensor would be both difficult and expensive, a quick-fix was found in allowing the spacecraft to coast over the South Pole in any attitude it chose (after all, there were no targets in Antarctica) and its proper attitude was restored as it entered warmer latitudes.

From May through October 1964, six GAMBIT flights produced coverage on only ▓ targets. Half of the flights produced no coverage whatever. On ▓▓▓ coverages delivered, the best resolution was seven feet. The users of intelligence had seen enough 2.5-foot resolution to regard poorer performance with barely-concealed contempt; they clamored for an end to uncertainties in product quality.

The first faint sign of relief appeared in December 1964, with the flight of GAMBIT No. 14. Although battery overheating had shortened the spacecraft's lifetime to one day and only ▓ targets were covered, the 2.1-foot (best) resolution gave promise for the future.

GAMBIT-1 Flight Summary—1964

G-1 No.	Date	Targets Covered	Best Resolution	Days on Orbit	Remarks
5	25 Feb 64		—	2.1	Profound yaw after rev 2
6	11 Mar 64		3.0 ft.	3.1	First flight of the stellar-index camera; first truly successful flight re target coverage
7	23 Apr 64		2.5	4.1	Two days at low orbit
8	19 May 64		2.0	1.0	Attitude control problems
9	6 Jul 64		50.0	0	Attitude control problems
10	14 Aug 64		7.0	1.0	Electrical/programmer problems
11	23 Sep 64		7.0	4.1	Focus error; gas leak
12	8 Oct 64		—	—	Agena failure; no orbit
13	23 Oct 64		—	4.1	Retrofire/re-entry problem; No recovery
14	4 Dec 64		2.1	1.0	Power supply problem; aborted on Rev 18

GAMBIT No. 15, launched in January 1965, improved the resolution value, once again, to 2.0 feet; the coverage of ▓ targets also set a record (although the majority of these coverages were not at the best resolution). Flight No. 16 in March 1965 set a fresh record for coverage (▓ targets); in April, this number increased to ▓ on flight No. 17; in May it peaked at ▓. Both flights achieved best resolutions of 2.0 feet and set record on-orbit times of 5.1 days.

When flight No. 18 in May 1965 turned in a performance equal to its predecessor, one could have thought that GAMBIT had finally moved from adolescence to maturity. But the 25 June flight of GAMBIT No. 19 dashed such hope: a massive short circuit cut the target coverage to zero.

SECRET
Handle via
BYEMAN-TALENT-KEYHOLE
Control Systems Jointly
BYE 140002-90

GAMBIT-1 Flight Summary—Jan–Jun 1965

G-1 No.	Date	Targets Covered	Best Resolution	Days on Orbit	Remarks
15	23 Jan 65		2.0 ft	4.1	Temperature control problems
16	12 Mar 65		2.4	4.1	
17	28 Apr 65		2.0	5.1	Diagnostic instrumentation added
18	27 May 65		2.0	5.1	
19	25 Jun 65		—	1.1	Massive short circuit

Changing the Guard

The summer of 1965 brought key personnel changes in the National Reconnaissance Office. Dr. Brockway McMillan, who had followed Charyk as director of the NRO, in 1963, was replaced by Dr. Alexander H. Flax, Assistant Secretary of the Air Force (R&D), on 1 October.[47] Maj. Gen. Robert Greer retired from the Air Force on 30 June, with the successful development and early operation of GAMBIT to his credit. He was replaced as Director of Special Projects, Office of the Secretary of the Air Force (SAFSP), by Brig. Gen. John L. Martin, Jr., who had previously been chief of the NRO Staff in the Pentagon and later, for one year, deputy to Greer at SAFSP. Colonel King continued in place as project director for GAMBIT.

NRO Director Alexander H. FLAX

USAF Col William G. KING

The most serious immediate problem facing Martin was whether GAMBIT No. 20 should continue to hold to a flight date of 9 July. After considerable study, he opted to follow the schedule inherited from Greer. Actually, the date slipped to 12 July, at which time Martin witnessed a comprehensive failure: the Atlas booster shut down prematurely and GAMBIT flew a 680-mile arc into the Pacific Ocean. Martin realized that the GAMBIT program, during those few minutes, had retreated to square one; his main task, as a new commander, was to diagnose and cure a seriously ailing satellite system.

The Philadelphia Story

As General Greer's deputy, Martin had absorbed a detailed knowledge of GAMBIT. His personal assessment of the record of 20 flights was that very dramatic failures—such as the one he had just witnessed—were usually the most easily corrected. The chronic, nagging failures seemed to be those based on strange little events which occurred quietly and just outside the reach of on-board telemetry. He also concluded that most of these subtle aberrations were clustered in what he termed "the trouble-plagued OCV."[48] Reviewing OCV performance, Martin grouped troubles under two headings: (1) unexpected loss of control gas, and (2) unexpected loss of programmer control. These losses, separately or in combination, transformed a healthy-looking GAMBIT into a zombie—a stupid creature circling the earth in unauthorized orbits, totally disinterested in attending to its assigned duties.

Years later, Martin could still recall the emotion of "watching a bird go dead" or "go gypsy." "You simply cannot imagine," he said, "the frustration you feel when, after establishing a clean orbit, and grinding out a few good operational revs, the bird reappears over the horizon with all control gas mysteriously expended, or with a deaf programmer."[49] He reminisced, further, that his experience with such frustration dated to the very beginnings of GAMBIT flight history: he had witnessed the agonies of early GAMBIT operations while visiting Vandenberg Air Force Base and the Satellite Test Center at Sunnyvale in 1963.

Martin recalled that when he and Colonel King had pressed General Electric representatives, at Flight Program Review meetings, for answers to OCV problems, the "answers" were often evasive and diversionary.[50] GAMBIT's control-gas valves were made at GE's Utica Plant and integrated into the OCV at GE's Philadelphia facility (later at GE Valley Forge). GE representatives attending Program Review meetings were (properly) GE Philadelphia people who (improperly) did not seem to be well-acquainted with the production and qualification-testing history of GE Utica valves. They maintained a certain aloofness and serenity in the face of adversity, occasionally observing that "Valves will be valves: sometimes they fail to close . . . ." or "Isn't this possibly the kind of problem one *has to expect* in a novel technological venture like space flight?"[51]

Keenly dissatisfied with such responses, King asked the GE Philadelphia representatives to accompany him to Utica, where they might observe, first-hand, the manufacture, assembly, and qualification testing of "new venture"

SECRET
Handle via
BYEMAN-TALENT-KEYHOLE
Control Systems Jointly
BYE 140002-90

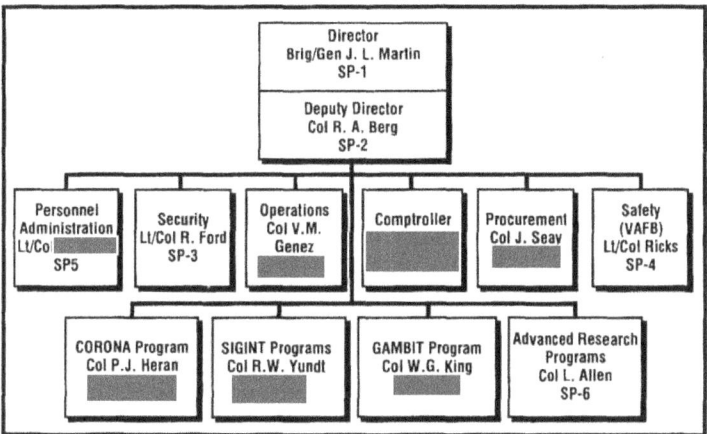

SAFSP Organization in 1965

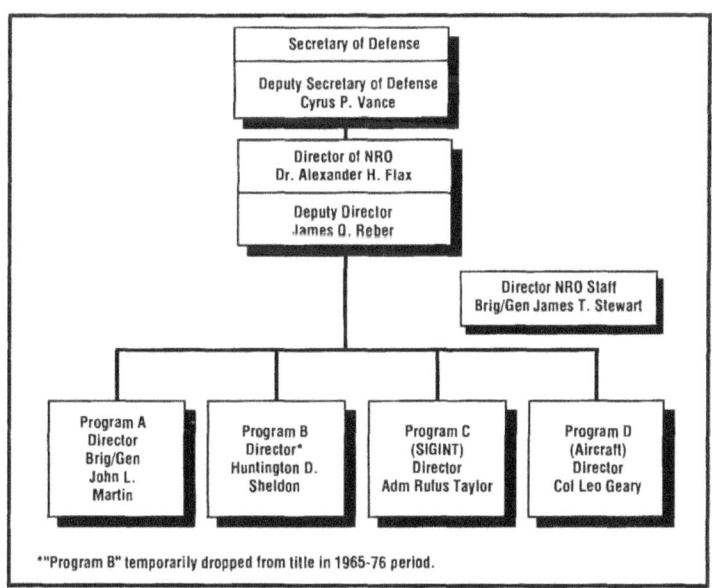

*"Program B" temporarily dropped from title in 1965-76 period.

NRO Organization in 1965

valves. It took less than a day to learn that the fault did not lie in design, manufacture, or assembly; the damage was occurring in the testing program itself—a process which, as applied by Utica personnel, was *inherently* capable of randomly damaging valves and making them liable to failure on orbit. As soon as the "lethal" elements were dropped from the testing procedure, the GAMBIT's control-gas problem subsided.

The second miscreant—the temperamental GAMBIT programmer—more than filled the anxiety gap previously occupied by faulty gas valves. The GAMBIT programmer was designed to be the first truly sophisticated "brain" on a reconnaissance satellite; without perfect performance by the programmer, GAMBIT capability was reduced severely—sometimes to the level of helplessness. In visits to GE Philadelphia, Martin and King had been faced by an attitudinal setting similar to that at gas-valve conferences. There was *deja vu* in knowledge that the programmer was made by GE Utica and accepted for integration into the OCV by GE Philadelphia.

Shortly after assuming command, Martin traveled to Philadelphia and requested an occasion to address all employees working on the OCV. He gave a short speech, specifying the problem clearly and describing, in some detail, the probable consequences of continued programmer failure. A few days after his return to Los Angeles, he was telephoned for an appointment by a GE Philadelphia employee, calling from the Los Angeles airport. Martin invited the gentleman to his office and learned, to his surprise and amusement, that he was the "company psychologist" for GE Philadelphia.

"I was present for your speech, General, and I think you should know about a serious misunderstanding as to exactly what you were saying to us, or, perhaps, by what you *meant* to say to us. Some of our people came away from the meeting with the impression—I hope we can clarify it—that unless work performance improved, they would be fired!"

The psychologist paused to let the full horror of this thought take effect. Martin's reply was cheerful: "Exactly! Go back and tell them that *they have the message.*"[52] It was a short meeting. From the earliest days of satellite reconnaissance operations, it was commonplace for customers to remark on the sharp divergence in response style of different contractors attending Program Review meetings. CORONA's flight operation had begun with 12 consecutive failures, providing ample opportunity for forming opinions and conclusions in this regard. GE Philadelphia's position at these meetings, absent specific data, was often that "It really couldn't have been *our* equipment that caused the problem." And that was that. The LMSC approach was somewhat different. After the routine and obligatory initial announcement that "the Thor was undoubtedly out of spec," LMSC's engineers would concede their own problem and settle down to solving it. Hard-driving Fred O'Green would open this part of the discussion with a standard exhortation, "Let's have no shyness here. Don't tell me how perfect your equipment is; tell me how it *could* have caused *this failure!*"

Different styles in confronting adversity seemed to be more a reflection of organizational arrangement than personnel bias. Lockheed, for example, habitually organized on a project basis—automatically selecting one responsible person and having him collect an engineering, administrative, and procurement staff, wholly responsible to himself. GE, on the other hand, was in the habit of organizing functionally. The GE project leader stood almost alone, channeling his needs to already-existing engineering, manufacturing, quality, test, administrative, and procurement divisions. As a result, persons working on his program in those divisions were not under his direct control; therefore, his daily *modus* called for diplomatic cajolery, rather than command. The contrast between the GE and Lockheed approaches to project organization could not have been more diverse.

Aware of the GE management style, and the extreme difficulty that would be encountered in amending it, Martin and King concluded that friendly persuasion would not work in contacts with Philadelphia. The conclusion may have become definitive on the day they heard, "We are trying our best," and "You simply have to expect some problems," capped by "If you'll tell us what to do, we'll be glad to do it"—all symptoms of a contractor reaching the end of his tether. Martin and King's preferred reaction would have been to cancel the GE Philadelphia contract, but this would have been self-defeating: it was far too late in the GAMBIT program to begin competitive work with a new contractor—such an action would delay the flight program for several years. Clearly GE and the NRO were "stuck" with each other and desperate measures were needed to deal with a desperate situation.

By September 1965, King was outlining a suitably desperate measure to General Martin. Shortly thereafter, they both proceeded to GE Philadelphia, where they mystified management by requesting exclusive use of a dining room, ten tables, ten white tablecloths, and ten completed GAMBIT electronic boxes. When everything was in place, Martin produced his own screwdriver and removed the cover-plates from a box. He then raised the box above the cloth-covered table and shook it as hard as he could. He paused to inventory the native and foreign items which had fallen to the cloth, identifying them in a disturbingly quiet tone. The chant continued as he and King moved from table to table, repeating the operation with each box, the findings confirming King's estimate of the quality integrity of this equipment.

Concluding the exercise, Martin observed to his hosts that *someone* (or "someones") had to be responsible for the debris on the ten tables. He repeated his view that although the boxes had been fabricated at GE Utica, GE Philadelphia had final responsibility for their condition and performance.

By late afternoon, GE Philadelpha management had identified a group of "someones" and announced a list of dramatic personnel actions. At GE Utica, both organization and procedures were revamped. A more experienced program manager was appointed, who reported to General Martin, in person, in Los Angeles, once each week for several months, explaining progress on a "get-well-Utica" plan. Utica made substantial changes in several components, in most of its production methodology, and—most important of all—in testing procedures. System-level tests would no longer be the screening point for faulty components or workmanship; checking would henceforth take place when components were built. System-level testing was to serve only as *confirmation* that all components had been manufactured and assembled properly. This was the first important step in moving GE hardware toward respectability in the GAMBIT components family.

Several other forceful actions were taken *solus* by King, who disciplined errant contractors fearlessly, whenever he judged hardware quality to be at risk. At Lockheed, for example, he noted a tendency to "cover" the fact that an Agena was "a little off schedule" by shipping it to Vandenberg *on schedule*, trusting that with a little more time (and perhaps some luck) the engineers would have a chance to complete their work in Vandenberg's Missile Assembly Building (MAB). In a program like GAMBIT, with its complicated technology and demanding flight schedule, such a maneuver was bound to be attempted: the pressures on the development team would force it to search imaginatively for every minute it could find, *wherever* it could find it. On the other hand, once such a habit (reliance on MAB-time) was formed, it could quickly develop into a regular procedure and a contractor's "right."

Recognizing the strategem and the danger, King expressed himself explicitly in opposing—and forbidding—it. He became even more concerned when he found a second major contractor—Eastman Kodak—tending toward the practice. When he confronted EK, stating openly that "unready" optical systems were being shipped to the MAB, Messrs Waggerhauser and Simmons protested formally to Martin, urging that King be censured for an "unacceptable attitude" toward their optics and their engineers. King countered with a draconian solution: he advised General Ritland, SSD commander, that he would, if necessary, close the MAB, *forcing* all contractors to deliver flight-ready hardware to the launching site.

This was the ultimate move King could have employed to convince everyone, once and for all, of his unswerving determination to guarantee hardware integrity. As for the criticism of his "unacceptable attitude," he shrugged it off with a characteristic twinkle. He had made his point.

On-Schedule or Over-Target?

For most of its lifetime—even in its Army Air Corps days—the US Air Force's main development and procurement experience had centered on aircraft acquisition. The management headquarters for this work was Wright Field—later known as the Wright Air Development Center and then renamed the Aeronautical Systems Division. At this installation, near Dayton, Ohio, the aircraft acquisition task was assigned to program offices; each office managing the research, development, and testing phases of an individual aircraft.

There were three conditions under which an aircraft program office could fall from grace in the eyes of the Air Force (and the US Congress): (1) the program's development cost could grow to more than had been appropriated to it (a "cost overrun"), (2) the program could fail to meet its advertised delivery-date ("schedule overrun"), or (3) it could fail to meet its performance specifications.

Cost overruns were a painful experience: they triggered serious criticism of the Air Force (and the Department of Defense) by Congress; they often led to very unpopular dollar "raids" on contemporary Air Force projects; they could even cause cancellation of the offending development. The Air Force had in-depth experience with each of these responses and tried to prevent them, insofar as possible, by supervising contractors closely and offering incentives for "coming in under cost."

"On-time delivery" was not as sacrosanct as "under-cost," but, over the long run, schedule slippages could destroy a project, particularly in areas where technology was evolving rapidly. In this regard, one of the most disturbing analyses produced during the mid-1950s was Thomas G. Belden's survey of 100 key projects at Wright Field, showing that 85 slipped 0.45 year or more per year, 22 (of that 85) slipped one year or more per year, and median slippage was 0.7 year per year.[53] The Air Force was properly alarmed by these numbers and began to reward contractors for "coming in ahead of schedule," in addition to showing good cost-control performance.

The third problem—that of meeting performance specifications—was not nearly as open to public view as money and schedule problems. In fact, one way to "ameliorate" a schedule slippage on an aircraft development was to accept the aircraft from the contractor well before it had demonstrated specified performance and then to identify the inadequacies as part of the in-house flight test program at Edwards Air Force Base. Brig. Gen. Charles Yeager describes this process:

> By definition, a prototype [aircraft] was an unproven, imperfect machine. It was usually underpowered, had controls that were too light or too heavy, new hydraulic or electrical systems that were bound to fail, and more than a few idiosyncracies . . . . Some defects were obvious: Convair's Delta Dagger was completely redesigned following the poor performance of its prototype. But other problems, like an unexpected vicious pitch up at high speeds or a dangerous yawing tendency, might be discovered late in a program, only after hundreds of hours of flying time. The test pilot's job was to discover all the flaws, all the potential killers . . . Testing was lengthy and complicated, resulting in hundreds of major and minor changes before an airplane was accepted into the Air Force's inventory.[54]

It was natural for General Greer's original GAMBIT staff—particulary his procurement officers—to follow the incentive patterns developed at Wright Field. As a result, the earliest arrangement with SAFSP contractors was weighted heavily toward rewarding cost underruns: the incentive for good performance on-orbit was only one-half the possible bonus for lowered cost.

Unfortunately, experience soon showed that the conjectured parallel between aircraft and spacecraft development programs was false. The programs were *fundamentally* different with the difference stemming from one hard fact: satellite reconnaissance systems had no Edwards Air Force Base.

SECRET
NOFORN-ORCON

There was no way to provide to GAMBIT the "hundreds of major and minor changes" that Yeager's flight-test program provided to aircraft prototypes. In addition, the Intelligence Community had great expectations for *every* GAMBIT flight—*even the first one*. Finally, GAMBIT, like any spacecraft, could not be taxied back into a hangar for modification or refurbishing. The intelligence environment made every GAMBIT flight a *mission*; the aircraft parallel would have been to move a newly-developed fighter directly from factory to combat. It was clear that the Intelligence Community—the customer—thought the news of a GAMBIT being under-cost was *interesting*; the announcement that it was ahead of schedule was *praiseworthy*; but the fact that it was performing correctly over target was *mandatory*.

During his service as General Greer's deputy, Martin had made an exhaustive study of incentive contracting and had arrived at conclusions which paralleled closely the views of the Intelligence Community. He believed that the problems that arose in the GAMBIT flight program were not being considered properly in incentive payments to contractors. He could not agree with a system of rewards in which the bonus for good performance on orbit was only one-half that paid for delivery under-cost. He observed, for example, that such a set of values placed General Electric in position to collect a healthy bonus (thanks to being under-cost) on a system which might fail to produce *anything* useful on orbit. The incentive system unintentionally hinted to the contractor that de-emphasis on quality control and testing procedures could actually increase the award for delivery under-cost.

Martin's "fix" was to de-emphasize the cost underrun bonus ("an inexpensive system that doesn't work is too expensive for us"), leave the delivery time bonus about where it was ("I *do* have customer pressures on scheduling"), but place extraordinary emphasis on orbital performance. His shift in emphasis was measured by a comparison of *bonus maxima* for good on-orbit performance: ▮▮▮▮▮ under the old system versus ▮▮▮▮▮ under the Martin plan. In fact, even with a cost overrun of 25 percent, perfect GAMBIT performance on orbit could win a bonus of ▮▮▮▮▮ for the contractor.[55]

The final novelty in Martin's plan was a touch of true genius: he decided to pay the *maximum* performance bonus to the contractor *in advance* of the flight; post-flight the contractor would write a check of his own, returning any of the on-orbit performance bonus he had failed to earn. Writing checks to the US Air Force was a disciplinary experience no contractor liked to envision, or, even worse, have cause to remember. In General Martin's words:

> My dominant concern about incentive contracts was that the contract's structure and elements did not penetrate to the level where success or failure is formed: it was simply a set of details known to the contracting office, not a concern of the engineers, technicians, and fabricators. I wanted an incentive structure that would result in people at these latter levels willingly doing something that they wouldn't have done without it. Payment in advance, with repayment required if not fully earned, was my approach to (a) getting the word to

SECRET
Handle via
BYEMAN-TALENT-KEYHOLE
Control Systems Jointly
BYE 140002-90

these levels, as well as (b) insuring that high management levels would be personally concerned with what their engineers, technicians, and fabricators were doing.[56]

Out of the Valley

The Martin-King drive for improvement in the GAMBIT program was unrelenting. In preparation for flight No. 21, the entire system was tested exhaustively and inspection procedures were upgraded impressively. All GE Philadelphia OCV components underwent X-ray examination, together with strong vibrational tests. Faulty components *were* discovered—and replaced.

The flight of No. 21, launched on 3 August 1965, hardly reflected the level of detailed attention this GAMBIT had received: the satellite lost stability almost immediately, as a power converter failed, making the satellite useless on orbit. After three bad flights in a row, the Intelligence Community now became very restless: an intensive ICBM buildup was going on in the Soviet Union and there had been no high-resolution coverage since May. Intelligence needs had previously been geared to one *successful* GAMBIT flight every 40 days.

GAMBIT-1 Flight Summary—1965-67

G-1 No.	Date	Targets Covered	Best Resolution	Days on Orbit	Remarks
20	12 Jul 65		—	0	Atlas failure. No orbit.
21	3 Aug 65		—	4.1	Power converter failure. No stability.
22	30 Sep 65			4.1	
23	8 Nov 65			1.1	Loss of stabilization gas.
24	19 Jan 66			5.1	First successful use of color
25	15 Feb 66			5.2	Stereo mirror servo-motor problem
26	18 Mar 66			6.1	
27	19 Apr 66			6.1	
28	14 May 66			6.1	First successful night photography. Street light patterns detected.
29	3 Jun 66			6.1	
30	12 Jul 66			8.1	
31	16 Aug 66			8.1	
32	16 Sep 66			7.1	
33	12 Oct 66			8.1	
34	2 Nov 66			7.1	Pyrotechnic/door problem; no camera operation
35	5 Dec 66			8.1	
36	2 Feb 67			8.1	
37	22 May 67			8.1	
38	4 Jun 67			8.1	

Fortunately, CORONA had been reasonably successful during the summer, but its photographic resolution capability, even at best, was not detailed enough to meet the expectations, or many of the needs, of the analysts.

Martin, characteristically, stood steady amidst this flurry. He and King found themselves, as usual, going against the flow, even proposing a one-month slip in the launching date of GAMBIT No. 22, to accommodate increasing comprehensiveness in testing and qualifying the components used by GE, where all modified components were now going through mandatory tests. Thermal vacuum tests and vibration tests were still uncovering faulty parts and assemblies. Contamination and workmanship problems were still surfacing, even though GE inspection teams were "bird-dogging" over 25 critical components in each OCV. Through it all, or perhaps because of it all, Martin was feeling an increased assurance that nothing was wrong with GAMBIT's basic *design*; he believed, based on detailed study, that the OCV could be, and was being, reformed into a "good bird."

GAMBIT No. 23 was the first satellite to have full benefit of the new test and inspection regimen at Philadelphia. It was also the first to have sufficient electrical power, as well as enough control gas, for a six-day on-orbit capability. Launched on 8 November, it quickly succumbed to the familiar, fatal flaw: excessive use of stabilization gas. During its 18-revolution lifetime it photographed only ▓ targets.

But the sun would soon shine: 14 of the 15 remaining GAMBIT-1 flights were rated "very successful"—even by the stringent standards of the Intelligence Community—averaging 6.6 days on orbit, ▓▓▓ targets covered per flight, and best resolutions ▓▓▓▓▓ 2.0 feet. Flight No. 30 was the first to spend more than eight working days on orbit. Flight No. 38 ended the GAMBIT-1 program, celebrating ▓▓▓ targets covered and a best resolution of ▓ feet.

Reviewing the GAMBIT situation of 1965, it is clear that redemption of the OCV—and its contractor—was the critical step in moving the program out of shadow into the sunlight. The new management team at GE worked hard—very hard—to cure a faltering OCV. The "Martin Specialized Incentive Contract Structure for Satellite Projects" continued to guide GAMBIT procurement practices and, together with corporate pride, furnished basic motivation for on-orbit success. The "Martin Motivator" earned the highest regard of both governmental and industrial officials; even today, it is frequently cited as the major positive influence in creating an extraordinarily successful (post-1965) GAMBIT program.

Section 4

Origins of GAMBIT-3

In mid-1963, about the time of the first GAMBIT launching, several reasons developed for SAFSP and the NRO seriously to consider improving GAMBIT's photographic capability. A basic motivation was the ever-increasing importance of high-resolution photography to the Intelligence Community and, concomitantly, the realization by intelligence analysts of the importance of even better resolution than that expected from the original GAMBIT. The pressures for better quality and greater quantity also resulted from engineering considerations: from a photographic payload point-of-view, the current GAMBIT configuration was non-optimal. It was non-optimal in a number of ways: redundant structural elements, thermal-management subsystems, and power-distribution gear consumed more than their fair share of allowable space and weight. These unnecessary duplications were crowding out possible growth elements: larger optics, more film, and more life-extending expendables.

In SAFSP-6, Navy Capt. Frank B. Gorman and his Advanced Developments Section had studied various improved payload and spacecraft configurations, working in concert with Eastman Kodak. Kodak also received strong encouragement to perform these studies from Col. ▬▬▬▬▬▬▬▬ who was in charge of GAMBIT procurement for SAFSP, and who believed that a more optimal design would be a major contribution to improved intelligence collection. The study goal was to determine the best resolution obtainable from an Atlas-Agena-launched system (the Titan-IIIB was later substituted for the Atlas). From studies such as these came the conclusion that an Agena-like stage, with a roll-joint, plus a more capable payload (larger optics and film load), could do the job without the then-current redundancy of the GE OCV and the Kodak payload. Roll-joint development had been kept alive at Lockheed by SAFSP and its technology was ready for application to an advanced version of GAMBIT. The roll-joint would allow the camera to be rotated to various positions in a plane perpendicular to the line-of-flight, permitting photography to the right and left of the ground track of the satellite. In addition to allowing the camera line-of-sight to be rotated, the roll-joint, by means of a carefully balanced internal flywheel, moved at a predetermined rate to offset precisely the inertia forces generated by rotating the forward (photographic payload) section of the spacecraft. It ultimately became a sophisticated device, with effective redundancy for high reliability. It was even able to compensate for a payload section whose mass (and thus its rotational inertia) would differ with, first, two recovery vehicles, and then one, the first having received its film-load and returned to earth.

Kodak Proposals for an Advanced GAMBIT

Beginning in 1961, several contractors were participating in studies of larger optics for reconnaissance purposes. Among them, Eastman Kodak had two advantages: first, it had designed and built the original GAMBIT system optics, which had produced excellent results; second, it had won an SAFSP-sponsored competition called VALLEY and, under that contract, had designed and built some key optical and film-handling components of a system which had as its aim large-area coverage at significantly better resolution than that achievable by CORONA. These advantages were supported further by an unwritten agreement between senior Air Force officials and Kodak management that committed the Air Force to sparing Kodak, as far as possible, from the "feast or famine" aspects of government contracting, in exchange for Kodak's willingness to divert some of its best people from more lucrative commercial work to supporting government needs.[57]

In 1963, Kodak employees Charles P. Spoelhof and James Mahar studied various GAMBIT improvement schemes. In December, they presented their results to DNRO McMillan, and, subsequently, to Greer. Following this, Greer formally proposed to the DNRO the development of an improved, higher-resolution GAMBIT system.

Fundamental to the Kodak studies was a rejection of the initial GAMBIT scheme, wherein the payload was carried inside an OCV. Kodak preferred two physically distinct modules, one containing the camera subsystem together with the recovery vehicle, the other housing propulsion and on-orbit control subsystems. These completely tested and flight-ready modules would be mated (at the pad) to the booster, using the "factory-to-pad" concept,

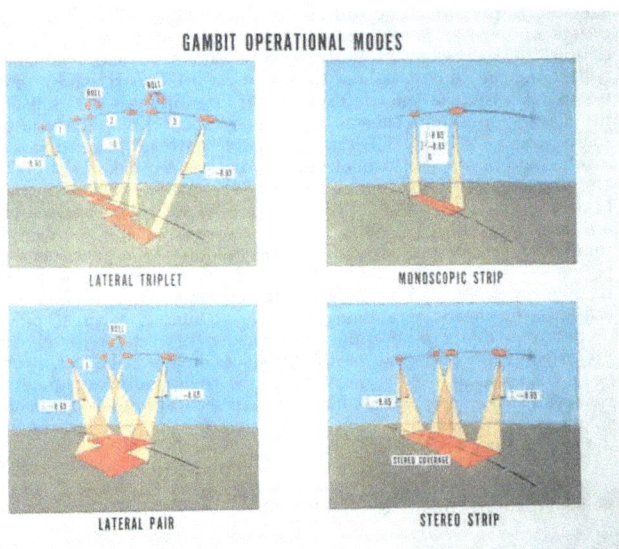

GAMBIT-3 Pointing and Stereo Capability

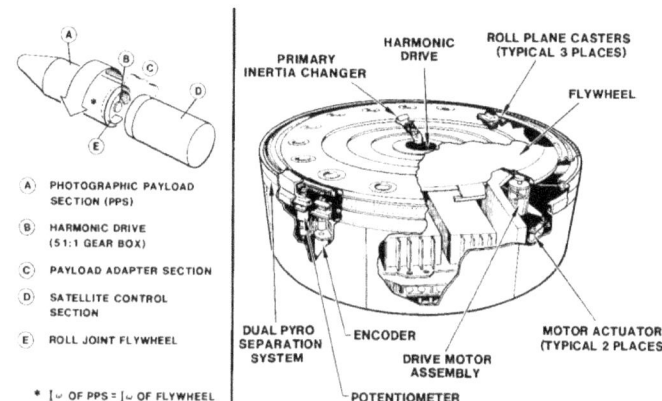

PAYLOAD ADAPTER SECTION
(ROLL JOINT)

GAMBIT-3 Roll-Joint (Payload Adapter Section)

obviating the significant assembly and test problems experienced at Vandenberg AFB throughout the initial GAMBIT program. In developing this approach, Kodak exploited use of the Lockheed roll-joint, which it planned to locate between the forward photographic payload/recovery vehicle section and the rearward satellite-control section.

Various degrees of improvement were studied. In each case it was assumed necessary to achieve the desired resolution 95 percent of the time, at a ▓▓▓▓ from an altitude of ▓▓▓▓. In these terms, a Kodak-conceived GAMBIT-2 would deliver an ▓▓▓▓ resolution; its GAMBIT-3 would have a ▓▓▓▓ capability; and its GAMBIT-4 would go to ▓▓▓▓. In each case, the variable was the aperture and focal length of the optics (all being catadioptric Maksutov-type lenses). Use of a special, ▓▓▓▓ for both the optical barrel and related assemblies was planned, along with a new thin-base (1.5 mil) high-resolution film, with an aerial exposure index of ▓▓ (roughly a factor ▓▓▓▓ over the film then in use on GAMBIT). These technologies had been proven and effectively used in the improved GAMBIT program.

In evaluating the possible advanced GAMBITs (GAMBIT-2, -3, and -4), SAFSP ruled out GAMBIT-2, since its promised ▓▓▓▓ resolution was not a significant advance over GAMBIT-1. As to GAMBIT-4, SAFSP thought the optics were so large as to create manufacturing uncertainties, long development times, and rather large costs. The NRO's view was that GAMBIT-3 should be favored, since it would provide significant improvements in resolution and target quantity at acceptable estimated costs and lead times.

DNRO McMillan was concerned that the program might have major problems if the larger optics and the improved film did not become available on schedule. As a hedge, he sponsored alternative optical materials developments. One was a material called ▓▓▓▓▓▓–a proprietary low-coefficient-of-expansion, glass-like material made by ▓▓▓▓▓▓▓; another was ▓▓ ▓▓▓▓▓▓ glass, also made by ▓▓▓▓▓ a third was an ▓▓▓ ▓▓▓▓▓ proprietary formulation called ▓▓▓▓▓ –a low-coefficient-of-expansion glass. Even Invar metal was studied as possible mirror material but it was found that ▓▓ had sufficient microcreep to make it unstable for large, high-precision optics. In addition, Kodak was authorized to work on various ▓▓▓ Despite these alternative technological endeavors, the final GAMBIT-3 mirrors were made to the baseline design: fused silica ▓▓▓▓▓▓▓▓ fused silica bottom and surface plates. Early mirrors were "tuned," using vacuum deposition of silica in selective areas. In later production, the necessary quality was obtained by conventional polishing. This final choice was not formalized, however, until late in 1965. New technologies did contribute to the improvement of large optics manufacturing with laser-interferometric testing, which ultimately became machine-readable and, with mini-computer support, provided quick turnaround from testing to polishing. Using these procedures, GAMBIT-3 optics consistently met or bettered the error-design goal (less than $\lambda/50$ peak-to-peak).

The NRO Selects GAMBIT-3

By 13 December 1963, Colonel King had prepared an initial development proposal based on the technical content of Kodak's GAMBIT-3 briefing. He incorporated Greer's instructions that general cleanup of the GAMBIT-1 system would continue until GAMBIT-3 became operational.[58] King's plan provided that the entire GAMBIT program (both GAMBIT-1 and GAMBIT-3) would operate under the purview of the existing SAFSP GAMBIT office. The initial flight of the new system was scheduled for the second quarter of 1966, with the operational transition from old to new GAMBIT planned for later that year. Contractors for the new system would begin "controlled entry" into development toward the end of FY-64.

King saw only two acceptable options for GAMBIT-3 orbital control. He conceded that an OCV could be developed with a capability similar to that of GAMBIT-1, but he favored using a flight-control stage—such as the Agena—with a roll-joint coupling to the photographic payload. The roll-joint, invented for LANYARD and adapted for the first few GAMBIT-1 flights, had operated perfectly, while General Electric's orbital-control vehicle, a new and complex system, had, as noted earlier, experienced many difficulties. To rely on the Agena for orbital control would offer advantages of lowered technological and financial risk.

King's tentative schedule called for receipt of proposals in mid-April 1964 and contract award by June. As 1963 ended, DNRO McMillan informally authorized Greer to proceed with the GAMBIT-3 program outlined in King'

~~SECRET~~
~~NOFORN ORCON~~

development plan; formal approval arrived at SAFSP on 3 January 1964. Although King and Greer were predisposed toward the space booster combination of Atlas and Agena, McMillan felt that the option of using a Titan-III should be retained (a judgment which later proved to be fortuitous), provided scheduling difficulties would not result. This consideration represented the only significant alteration of King's initial GAMBIT-3 plan.[59]

King and Greer worked out the remaining major elements of the GAMBIT-3 concept early in January 1964. The major subsystems of the satellite were to include a photographic-payload section (PPS), a satellite-control section (SCS), and the booster vehicle. The payload section would contain a camera module and a satellite recovery-vehicle (SRV). The control section was to include the command system, the orbit-adjust module, an attitude-control subsystem, a back-up stabilization system (BUSS), and the power supply. Although the option of using a GAMBIT-1-style orbital-control vehicle had not yet been formally discarded, as early as January 1964 the Greer-King concept leaned heavily toward Kodak's proposed approach.[60]

By 3 February 1964, the Program Office had prepared a General Systems Specification, a Satellite-Control Section (SCS) Work Statement, and a Preliminary Development Plan, together with a determination of sources to be solicited for the new program. By the next day, it had completed the SCS Request for Proposal, which was approved on 5 February. The competition was, of course, for the SCS job and the two bidders were the "incumbent" GE, which had built the GAMBIT-1 OCV, and Lockheed, which had built the Agena and the roll-joint. Kodak would be the PPS contractor.

It was at this time that the Program Office's documentation began to call this advanced GAMBIT by the name G³ or GAMBIT-Cubed instead of GAMBIT-3; for reasons of consistency, this history will refer to the program only as GAMBIT-3. In the Preliminary Development Plan the objectives of program were:

- High reliability;
- Flight capability commencing 1 July 1966;
- Ground resolution of ▮▮▮▮▮▮▮ contrast, as presented to the front aperture of the camera in black-and-white film for vertical daylight photography at a payload altitude of ▮▮▮▮
- Day or night photographic coverage of targets. Exposure commandable on orbit from a selection of exposure steps;
- Use of black-and-white film, color film, and/or other special recording materials;
- Stereo, strip, and lateral pair photography;
- Dual-camera subsystem configuration;
- Minimum mission life of eight days;
- Utilization of Atlas/Agena-D launching vehicle, or equivalent;
- Direct factory-to-pad cycle;
- Air recovery over water (primary) or water recovery (emergency).

The development plan emphasized maximum usefulness of the final photographic product by including user aids, such as optimized time-track format and a means of recording altitude data. It reduced operational constraints, such as environmental-door operation and film-drive transients. It described problems with the GAMBIT-1 design in "that testing and maintenance were greatly hampered by the integrated nature of the subsystem such that a great deal of connecting and disconnecting of wiring harnesses was necessary. Components were very difficult to remove and replace." In the GAMBIT-3 system, "the emphasis on modular construction will increase the ease of maintainability."[61] The plan stressed that reliability should be enhanced "through use of previously qualified subsystems" and took the position that "the command-and-control subsystem developed under the G-program has been shown to be a superior system, and it is anticipated that it and associated software may be included in the G^3 System."[62] (That anticipation was realized and all command programmers for GAMBIT-3 were produced by GE Utica on a sole-source basis.)

In approaching GAMBIT-3, the Program Office took the view that the government would allow contractors to enter "system development under carefully controlled conditions with funds utilized only as necessary to further define the system, provide system analysis, demonstrate feasibility of critical components, and protect long-lead-time development of specified components and/or subsystems."[63] This cautious approach on a high-priority system was established so that "the System Program Office can judge progress and feasibility for continuing into complete system development."[64] Despite such conservatism (for a system scheduled to fly in less than two and one-half years), the program proceeded apace.

Eastman Kodak was placed under contract prior to the selection of the SCS contractor. By mid-May 1964, Lockheed had been selected as the primary SCS contractor; GE was given a backup SCS study. As previously noted, the GE command programmer had been selected and GE established as an associate contractor, providing equipment to Lockheed as government-furnished equipment (GFE).

Kodak chose GE as the source for the GAMBIT-3 recovery vehicle (RV) and chose Lockheed for external PPS structure and certain components (such as a cutter-sealer to be installed on the film-path between camera and RV).

Because of DNRO McMillan's strong interest in the Titan as a possible booster for GAMBIT-3, Lockheed was tasked, in July 1964, to study Agena compatibility with the Titan-III(X). GE was asked to study the simplification of its GAMBIT-3 proposal, including the possibility of using a separate ascent stage with the orbital-control vehicle to provide the SCS capability. Before long, progress at Lockheed was sufficient to permit SAFSP to fund Lockheed's SCS work only, ending the GE backup study.

In October 1964, the SAFSP staff prepared cost estimates pertaining to the switch from an Atlas-Agena to a Titan-III(X)-Agena, based upon the Lockheed study. There were several reasons for considering such a change. One was the desire to use the Titan-III family of boosters for other Air Force space missions, thus providing a more efficient production base and commonality of ground-support equipment; further, the Titan had potential versatility and on-orbit weight-growth capability; finally, although the Atlas was considered to be an Air Force standard launching vehicle, technical control and production contracts for the Atlas were seen as shifting to NASA. A final influence on Greer was the likelihood that, in the not-too-distant future, the CORONA program would be replaced by a larger, longer-lived search system, which would require a Titan-III booster. Just before the end of 1964, Dr. McMillan approved the GAMBIT change from Atlas to Titan and the Titan-III(X) was officially designated as the Titan-IIIB.

GAMBIT-3 Developments at Lockheed and Kodak

Lockheed's work on Agena modifications and the roll-joint proceeded without major difficulty and never was a threat to achieving a successful launching on the planned date (July 1966) at close to budgeted cost.

By the fall of 1964, Kodak and its subcontractors had progressed to where engineering specifications of the overall design had been released, preliminary design reviews held, some engineering drawings released, and some critical long-lead-time components ordered. Payload development, however, was behind schedule.[65]

The major problem at Kodak was the manufacture and mounting of the two large mirrors of GAMBIT-3 optics. The primary mirror ▓▓▓▓ was ▓▓▓▓▓▓▓ and the stereo mirror ▓▓▓▓▓▓▓ ▓▓▓▓ These optics were larger than those of many terrestrial telescopes, but were required to be much lighter, with optical figure accuracy at least as demanding. Kodak experienced several failures caused by collapsed and fractured substrates of the mirror blanks. In addition, the figuring and polishing processes in the development phase were requiring much more time (to get the desired accuracy) than had been expected. Kodak had originally estimated that each of the two mirrors would require about 800 hours of grinding, polishing, testing, and coating from raw blank to finished product. The manufacturing time for the early mirrors ran as high as 3,000 hours per mirror, although this number was substantially reduced in later production. These difficulties led both the government and Kodak to continue alternative material and processing method developments, as mentioned earlier.

During the last quarter of 1964, Kodak remained behind schedule, while all other parts of the program were on, or ahead of, schedule. To the GAMBIT-3 managers at SAFSP, this was a tolerable situation; they had expected more problems during the critical first year of the program. They were pleased that most of GAMBIT-3's technical problems had been identified and resolved, the only exception being the optics. At the end of 1964, the GAMBIT Program Office was able to return nearly ▓▓▓▓▓▓▓ in uncommitted funds to NRO budgeteers.

By late 1964, Eastman Kodak had progressed to where it had developed sound techniques for manufacturing the large optics. Despite the promise of unconventional techniques—such as the selective deposition of silica used in the first few GAMBIT-3 mirrors to improve optical figure—the large optics were to be manufactured by conventional means of polishing, but with

Numerical Summary of GAMBIT-3 Payload

Photographic Output Data
Ground Resolution (R-5 lens)
  (vertical photography)

Lens-film resolution
  (dynamic)
Scale of photography (90 nm altitude)
                             (65 nm altitude)

Width of photographed
  strip (90 nm-vertical)
Scene width on payload film[1]
Scene length on payload film   variable
Scene length on ground   variable
Number of photographs

Payload Dimensions
Weight (at launching)
  Photographic Payload Section   4,130.0 lbs
  Film
    Primary Camera   10,000 ft of UTB 9.5-in wide
    APTC[4]
      Terrain Camera   1,080 ft of 5-in
      Astro Camera   780 ft of 35-mm
  Dimensions of PPS
    Maximum diameter
    Length (nose to interface bolt circle)

Camera Optics Module (included in PPS)
Dimensions
  Length
  Diameter
  Weight
Primary Camera Lens
  Focal Length
  Clear aperture diameter
    (primary mirror)
  f/number

Type
  Catadioptric aspheric reflector with four-element Ross corrector
  Configuration line-of-sight passes through stereo mirror
    Semi-field angle
    Nominal Obstruction (depends upon stereo position)
    Nominal transmission
    Nominal T-stop
    Depth of Focus (Rayleigh Criterion)
  Lens tube of Invar w/coeff of thermal expansion

Production Mirrors
Material   ULE glass w/coeff of thermal expansion
  Primary   (Asphere)
      Figure: peak-to-valley
              RMS
      Weight (mounted)
      Size (round)
  Stereo   (Flat)
      Figure: peak-to-valley
              RMS
      Weight (mounted)
      Size (elliptical)
  Ross Corrector Elements
      Figure: peak-to-valley
      Weight
      Size

[1] On 9-inch film
[2] Based on looper film capacity of 60 inches maximum
[3] Effective flight No. 48 using dual-platen (9- and 5-inch) camera with 12,241 feet of 9-inch film and 3,874 inches of 5-inch film
[4] Astro Position Terrain Camera Table

Numerical Summary of GAMBIT-3 Payload (continued)

Strip Camera
Exposure time (nominal)

Slit width for nominal film drive rate
  No. of slits (for daylight)
  Range of slit widths
  Supplemental slits   Night slit
                               Test slit

Spectral band pass
Film-drive speed
  Nominal
  Range
  No. of steps
  Tolerances
Film Type

Astro-Position Terrain Camera (APTC)
Components
  Astro-position Camera (APC)
    APC Line-of-sight—
    above plane parallel to XY plane, one in +Y other in -Y direction
      Semi-field angle
      Format on film
      Film
      Shutter

      Nominal Exposure Time
      Performance

  Terrain Camera (TC)
    TC Line-of-sight—
    axis
      Semi-field angle
      Format on film
      Film

      Ground Format (90 nm vert)
      Stereo Overlap
      Shutter

      Exposure time

      Performance
        Static

        Dynamic

Satellite Recovery Vehicle

Number of SRVs:	
Vehicle Nos. 1 through 22	1 SRV
Vehicle Nos. 22 through end of program	2 SRVs
Weight (at ejection w/film)	
SRV No. 1	376.0 lbs
SRV No. 2	394.9 lbs
Length	42 in
Base Diameter	33 in
Dispersion	100 x 15 nm
Nominal Re-entry Angle	1.98°
Recovered Weight	
SRV No. 1	180.0 lbs
SRV No. 2	200.0 lbs
Parachute Deployment Altitude	35,000–40,000 ft
Time from Separation to	
Parachute Deployment	10 min
Time from Separation to Recovery	23 min

*Area Weighted Average Resolution

controllable figuring and polishing machines. There would be more rapid turnaround because of the speed and accuracy of laser-interferometry testing.[66]

Only one major change to the preliminary specifications occurred in the early stages of development. After consideration of risks, General Greer decided it would be imprudent to anticipate full development of a dual-platen camera configuration in time to meet flight schedules. DNRO McMillan agreed. (It should be noted that a dual-platen camera with both a nine-inch and a five-inch platen flew with great success in GAMBIT-3 flights 48 through 54.) Some changes in the photographic payload section occurred because of refined requirements. One of its components, the astro-position terrain camera (which was to determine the precise location of terrestrial features with respect to fixed stars) had been completed early by Kodak, with some redesign to enhance its performance.

Kodak had progressed in refining mounting methods for large optics so that on-orbit distortion would be avoided and, yet, heavy loads (during the flight's powered-ascent phase) could be withstood without consequential movement of the optics. The optics had to be held tightly, but not too tightly. In another major area of concern, Kodak continued to refine thermal-control techniques to preclude distortion of the large optics barrel by variations in solar energy input. As a result of these preventive measures, neither problem occurred in actual flight.

GE, under its Kodak subcontract, completed design of the GAMBIT-3 recovery vehicle (RV) in November 1964. Greer and King had insured, through Kodak, that the GAMBIT-3 RV would differ from the GAMBIT-1 RV only where absolutely essential. Kodak responded with an extremely stringent quality-control process.

At Lockheed, the GAMBIT-3 program was assigned to the Space Systems Division, under the direction of Program Manager Harold Huntley. In turn, Huntley reported to James W. Plummer, assistant general manager for Special Programs, and was supported by Robert M. Powell, chief systems engineer; John Harley, Design Engineering manager; and Robert Kueper, Controls (business) manager. At the time GAMBIT-3 work began, Agenas were produced by contract with the Air Force's Space Systems Division, using "white" contracts, as Standard Agenas (Agena-Ds). The Standard Agenas were then turned over to the "using program" to be "customized." In the case of GAMBIT-3, that process entailed removing standard Agena components not needed for the mission and adding those peculiar to the mission. In the front rack of the Agena (ahead of the large propellant tank), the principal added components were the attitude-control subsystem (which allowed the Agena to act as a precise and stable tripod for the GAMBIT-3 camera, as well as providing ascent guidance), the extended and minimum command subsystems (GFE from GE), many components of the tracking, telemetry, and command subsystem, the flight batteries (carried in both the roll-joint and forward rack) and power-distribution components, together with some parts of the back-up stabilization system, known more familiarly as BUSS or Lifeboat. The principal addition to the aft rack of the Agena was the secondary-propulsion subsystem used for orbital adjustment. Lifeboat components were also on the aft rack.

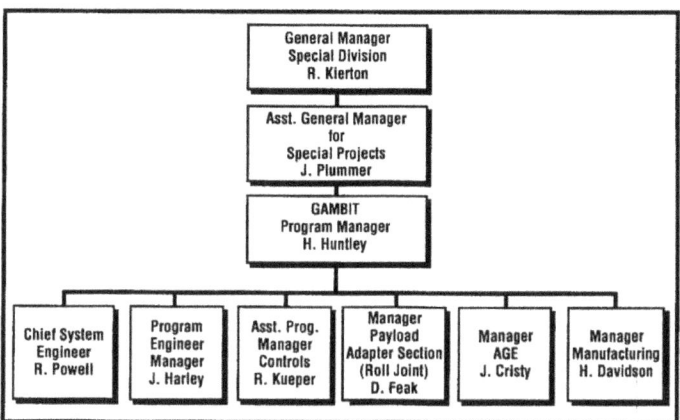

Lockheed GAMBIT-3 Program Office Organization

On a cost basis, some 35 percent of the components that went into the Agena were subcontractor-provided and, although there were some problems in development and manufacture, none was ever of a magnitude to pace the program.

The only SCS area where trouble developed was GE's GFE command programmer which, in March of 1965, was six weeks behind schedule. This slippage was offset by a schedule "glitch" at the Martin-Marietta Co., where the earliest possible date for a launchable Titan-IIIB booster turned out to be 28 July 1966—almost a month later than dates scheduled for the PPS and SCS contractors. As there appeared no way to protect the planned launching date of 1 July 1966, Greer proposed rescheduling to 28 July; DNRO MacMillan approved the change.

Experience with GAMBIT-1 had convinced the program office to incorporate a redundant viewport door actuator and backup film cutter on GAMBIT-3. At inception, GAMBIT-1 actuators were primarily pneumatic; these had failed in flight and were gradually replaced by electromechanical devices. Lessons learned were applied to GAMBIT-3 and most of its primary actuators were electromechanical.

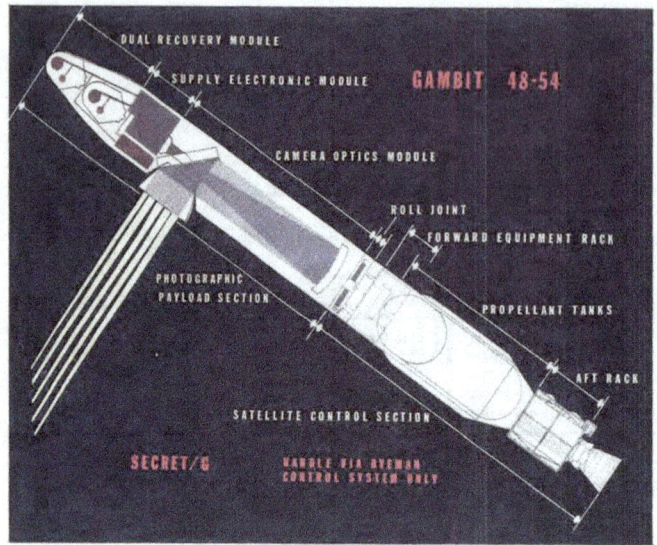

GAMBIT-3-Agena Vehicle

By the end of June 1965, GE's command system engineering was two months behind schedule. Because of mirror-fabrication problems, Kodak was three months behind. Kodak's problem was compounded by underestimation of the engineering manpower needed to produce electromechanical components for the PPS.[67] Fortunately, some schedule cushion had been built into the planned test spans, so neither delay affected the actual launching date.

During FY-65, GAMBIT-3 underran its budget estimate of ▓▓▓▓ by ▓▓▓▓. By anyone's standard, this reflected skillful cost estimating on a large, high-priority, advanced technology governmental program.

Technical problems, reflected in schedule delays, continued to nag both Kodak and Lockheed. Kodak's PPS schedules were so tight as to bring into question the 28 July 1966 launching date; doubt persisted through 1965. In some areas, Kodak was six months behind schedule; in September 1965, for example, testing of the PPS reliability model, which had been scheduled to begin on 15 October, slipped to April 1966. Schedule problems at Lockheed were less consequential and arose in connection with adding redundancy features to the highly important roll-joint.

Kodak was concerned over a shortage of technical people. In addition to GAMBIT-3, Kodak had a major role in three other NRO programs (including GAMBIT-1) and was providing the camera subsystem for NASA's Lunar Orbiter. Although Kodak was adding staff to meet these expanded needs, there was concern in SAFSP over the effect on GAMBIT-3 of Kodak's total load.

During the summer of 1965, problems on GAMBIT-1 were triggering other federal trauma: three successive failures had stopped the flow of high-resolution photography to intelligence users. General concerns prompted an intensification of quality control by all GAMBIT contractors. One component—the GE command programmer—used by Lockheed in the GAMBIT-3 SCS, received the most attention, followed by other elements of the command subsystem. Lockheed agreed (1) to acceptance test and inspect all command-system components, beginning with the hardware for GAMBIT-3 flight vehicle No. 6 and, (2) to accept the GE hardware within its performance incentive.

Because of the PPS schedule problem, a decision was made in early 1966 to reorganize the remaining test schedule for both the PPS and the SCS. If these new schedules could be maintained, GAMBIT-3 would make its planned 28 July 1966 launching date. They were and it did. A remaining concern was whether the new thin-base photographic film, with an aerial exposure index of 6, would be available for the first flight (it actually became available for the third flight).

Development test models of both the SCS and PPS were on schedule at Lockheed in April 1966, with completely satisfactory results. One feature of the GAMBIT-3 testing method that contributed to success of the spacecraft program was the use of computer-based testing, which sent test stimuli to the command subsystem, comparing the obtained response to the correct response. A similar testing procedure was used on the SCS and PPS at the Lockheed and Kodak factories, respectively. Computer-based testing of the SCS was unusually demanding. The entire SCS, the Agena and roll-joint, were put in a thermal vacuum chamber and run through all the phases of a simulated flight. These integrated tests were preceded by rigorous component-level acceptance tests, before the components were assembled into the SCS. The complete SCS vehicle tests, called "Programmed Integrated Acceptance Tests" (PIAT), were very thorough and fully automatic. The PIAT was run again at the launching base and definitely contributed to the subsequently successful flight program. The importance of the this concept is exemplified by the following quotation from SAFSP's GAMBIT-3 Status Book:

> The entire PIAT series of tests consist of a carefully programmed group of test sequences that thoroughly evaluate the health of each vehicle system as well as the interactions between the systems. The individual PIAT tests differ only in the areas where different environments (horizontal versus vertical, vacuum versus ambient, and mated versus unmated) preclude identical tests. The basic premise of concentrating the talents of the contractor and customer personnel in the development of a single comprehensive automated test (with variations) is believed better than a series of somewhat unrelated tests not conducive to automatic checkout.[68]

If a change was made in the flight vehicle anytime during the prelaunching sequence, the entire PIAT could be rerun to assure vehicle integrity.

GAMBIT-3 Launch Vehicle at Liftoff

Section 5

GAMBIT-3 Flight Program

By end of June 1966, it appeared likely that the planned launching date could be met. The Titan had arrived at Vandenberg AFB Space Launch Complex Four on 7 May; three days later it was mated to the SCS development test vehicle, with no interface anomalies. Preparations for the 28 July 1966 launching continued to go according to plan right up to the terminal countdown, but the attempt aborted at T-7 seconds, due to a ground-guidance equipment malfunction. Following necessary repairs, the count was resumed the next morning and, at 1130 PDT on 29 July, the first GAMBIT-3 mission, No. 4301, was successfully launched into orbit. The satellite achieved a near-nominal orbit and subsequent operations during the five-day mission went exactly as programmed. Compared to the PPS acceptance test results, which predicted a best-system-resolution capability of ▓▓▓▓ the KH-8 camera's demonstrated in-flight ground resolution of ▓▓▓▓ was more than gratifying and was about as much as could be expected from initial optical system flight hardware, according to the Performance Evaluation Team (PET) report. The report went on to state that for the first GAMBIT-3 flight "the intelligence content of this mission was the highest of any satellite mission to date, due primarily to the larger scale and better resolution of the camera system" and that "the main camera operated throughout the mission and the system acquired all targets as programmed."[69] This amounted to ▓▓▓▓ ▓▓▓▓ were successfully "read out."[70]

There were some PPS anomalies during the flight. Due to a slight main camera film misalignment, the time-track on the edge of the film was missing. The time-track was very important in the mensuration of strip film; its lack "caused considerable work in producing intelligence measurements from this system."[71] Another PPS problem resulted from an apparent shutter malfunction in the stellar camera part of the astro-position terrain camera (APTC) system so that "the Stellar Camera failed in almost every respect except on the dark side of the earth."[72] The terrain camera part of the APTC performed well with a nadir resolution of approximately 120 feet. The PET report noted that "the (terrain) camera provided some intelligence information to the community (this is the first extraction of intelligence information from an index-type camera)" and "the Terrain Camera provides an excellent mapping base to the mapping community with its improved resoluton over previous systems and with the increased scale and format."[73]

The photographic satellite vehicle (PSV) (including both the SCS and PPS) performed exceptionally well, as did the Titan booster. The satellite achieved a near-nominal orbit as follows:[74]

	Nominal	Actual
Inclination (degrees)	94.01	94.14
Period (minutes)	88.85	88.73
Apogee (nm)	160.00	152.00
Perigee (nm)	83.20	83.00
Eccentricity	0.0097	0.0090

On-orbit operations were nominal during the photographic portion of the mission. As this was the first flight, some operational constraints were imposed. One involved the allowed duration of operation of the roll-joint. The problem, which had surfaced at LMSC as a result of roll-joint qualification testing, involved the thermal margin of the roll-joint tachometer. The night before the Air Force's flight readiness meeting at Vandenberg AFB, James Plummer, Robert Powell, and Frederic Oder had hovered over a critical test that was being conducted at LMSC's roll-joint test laboratory in Sunnyvale, by Peter Ragusa, engineering manager for the roll-joint. The results of the test convinced them that a problem existed and, the following day, in a meeting with Martin and King at VAFB, they took the position that the PPS was not launch-ready without a constraint on the operating time of the roll-joint. After some discussion, General Martin agreed to restrict the roll-joint operations as follows:

Constraint	Mission Segment
180 sec ON during a 5,400-sec period	Rev 1 through 5
300 sec ON during a 5,400-sec period	Rev 6 through 10
180 sec ON during a 5,400-sec period	Rev 11 through 18
300 sec continuous ON and 450 sec total ON in 4,000-sec period	Rev 19 through 33
450 sec continuous ON and 450 sec total ON in 1,450-sec period	Rev 34 through 52
450 sec continuous ON and power OFF twice as long as ON	Rev 52 through recovery

The roll-joint did fail in the the post-photographic ("solo") phase during the 90th revolution; LMSC used an improved tachometer in subsequent missions. But for a first flight, the results were surprisingly good; even where problems appeared, the engineers were confident that they could supply definitive corrective action.

GAMBIT-1 or GAMBIT-3? How Many?

The initial flight success of GAMBIT-3 created a procurement problem for NRP planners. With two systems in being, each essentially capable of doing the surveillance mission, what should be the makeup of future procurements? Should GAMBIT-1 production be curtailed (or terminated) in light of this first mission? Or should it be maintained at some prudent level until GAMBIT-3 had fully demonstrated its development objectives? The problem was complicated further by the anticipation of a new search system named HEXAGON. The US Intelligence Board's endorsement of the HEXAGON/KH-9 system, in April 1966, made it appear advisable to begin near-term conversion of the GAMBIT-1 launching pad for later use by HEXAGON.

These concerns were overshadowed by a more immediate issue: should the NRO purchase all, or just a portion of, the 16 additional GAMBIT-1 systems originally scheduled? If one assumed a moderately successful GAMBIT-3, the 16 additional vehicles were redundant and their heavy cost an unnecessary burden. On the other hand, if GAMBIT-3 had serious problems in early usage, the availability of more GAMBIT-1 systems would be providential.

DNRO Alexander Flax stood on the conservative side of this issue, reluctant to cancel *any* GAMBIT-1 launchings until GAMBIT-3 had clearly proved itself as a viable follow-on system. By contrast, DCI Richard Helms argued that the total combined number of 20 GAMBITs (GAMBIT-1 and -3) in the FY-67 budget was excessive. He pointed out that the FY-67 schedule had been developed in the dark days of January 1966, reflecting a series of failures in late-1965. But there had been *no* recent GAMBIT failures; rather, impressive advances had been made in orbital lifetime and photo-coverage. It was Helms' opinion that the initial success of GAMBIT-3 was sufficiently compelling to warrant an optimistic outlook.

Hearing both sides of the issue on 17 August 1966, the NRP Executive Committee decided to delete four GAMBIT-1s from the buy-program. In addition, USIB's Committee on Overhead Reconnaissance (COMOR) proposed, in September 1966, an FY-67 flight schedule of nine GAMBIT-1s and eight GAMBIT-3s. USIB agreed with this proposal, members observing that even a moderate level of success, coupled with this scaled-down schedule, would result in saturating the user community with high-resolution photography. For the moment, then, the decision to proceed with a mix of GAMBIT-1 and -3 systems, during the 12 months starting in July 1966, was permitted to stand unchanged.[75]

GAMBIT-3 and the Needs of the Intelligence Community

By this time, the statement of Intelligence Community requirements for surveillance photography had become more detailed than previously and could be grouped under convenient headings of (1) better resolution, (2) more targets covered, (3) emergency response, and (4) timeliness. The GAMBIT-3 development team proposed to achieve "better resolution" by using newly improved optics and film. "More targets covered" was to result from extending orbital lifetime (which, from an engineering point of view, meant creating the capability to lift more expendables into orbit).

"Emergency response" was a more difficult problem, with some unrealistically expensive (possible) solutions; however, one could move in the direction of *shortening* response times. In August 1966, the NRP Executive Committee authorized extending the orbital life of GAMBIT-3 and modifying the satellite to carry two recovery vehicles. A longer-lived system could return its photographic "take"—upon demand—in recovery vehicle A, while additional "take" would feed into vehicle B for later delivery. (This GAMBIT capability was exercised for the first time in August 1969.) A second approach toward emergency response would have been to have a reconnaissance satellite on orbit at all times, that is, to have no "down-time" in the flight program. A third tactic might use a backup booster and PSV maintained in as near-to-launching condition as economically feasible. It was found that the cost of maintaining such a backup GAMBIT-3 system at a seven-days-from-ready-to-launch condition was not prohibitive. For most of the GAMBIT-3 program the capability was maintained, but never truly demonstrated. Although there were flight failures that caused the next-scheduled hardware delivery to be accelerated, the seven-day backup was never needed, as such.

The requirement for timeliness could only be fully satisfied by real-time readout. GAMBIT-3 could not move very far toward such a revolutionary goal; real-time readout, at the data rate desired, would require unique technology which was still under development in the 1960s, and would not be available for operational systems until the mid-1970s.

GAMBIT-3 and the GAMBIT-1 Heritage

The first GAMBIT-1 flight was launched in an atmosphere of tentativeness and speculation; by contrast, the first GAMBIT-3 launching was made with reasonable assurance. To the credit of the older system, three years of GAMBIT-1 experience had provided a technological heritage directly applicable to accomplishing GAMBIT-3 objectives.

High among GAMBIT-1 contributions was experience with orbital-control problems. Greer and Martin had seen enough of those episodes to be convinced that GE's orbital-control vehicle was, at best, generically temperamental. Prudence called for a fresh start in orbital control and Lockheed's Agena was the logical alternative, particularly since Agena was performing well on companion space programs and was being "standardized" by the AF Space Systems Division. The more King looked at Agena, the more he came to favor it (with roll-joint coupling to the photographic payload) as a preferred spacecraft.

Early GAMBIT-1 experience had demonstrated the need for a backup stabilization system on the spacecraft and one could be sure that the well-tested BUSS/Lifeboat would appear on all GAMBIT-3 vehicles. GAMBIT-3's horizon sensor would have separate profiles for winter and summer, thanks to experience over the Antarctic with GAMBIT-1. After years of proven performance in both CORONA and GAMBIT-1, the General Electric recovery vehicle would be a sound selection for GAMBIT-3.

Experience with GAMBIT-1 checkout procedures had shown, unequivocally, that a launching pad was not the optimal location for system checkout. Early in the GAMBIT-3 game, Greer, Martin, and King decided to require that checkout be done at Eastman Kodak and Lockheed. GAMBIT-3 would certainly be designed for automated checkout, during final assembly at those plants; then it would go directly to the launching pad, in accordance with the innovative factory-to-pad concept. Integrated subsystems had been used on GAMBIT-1; on GAMBIT-3 they would be modular.

Experience with quality control on GAMBIT-1, especially at GE Philadelphia, was fresh in the minds of SAFSP Project Officers; there would be special watchfulness on their part, particularly on a new program.

Finally, General Martin's incentive contract structure was in full effect and, accordingly, contractors' exertions and rewards would continue to be biased heavily toward excellent performance on orbit.

This legacy of experience in equipment, procedure, and methodology was unique in the history of satellite reconnaissance. It provided a solid basis for producing a new system capable of acquiring intelligence information on its

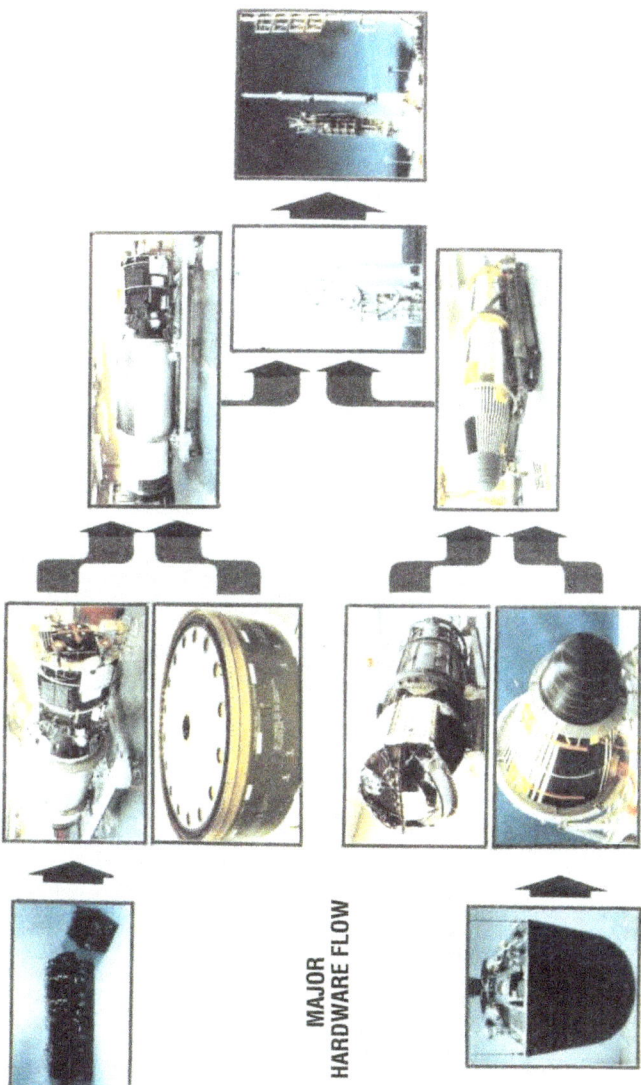

GAMBIT-3 'Factory-To-Pad' Concept

MAJOR HARDWARE FLOW

first flight. Later history showed that when defects *did* appear in GAMBIT-3, they were usually the product of oversight or accident, rather than a failure of process.

GAMBIT-3 Operations—The Flight Program

The GAMBIT-3 *development* flight program was limited to six vehicles, with these results:

GAMBIT-3 Flight Summary—1966-67

G-3 No.	Launching Date	Photographic Days	Best Resolution	Remarks
1	29 Jul 66	5		
2	28 Sep 66	7	36	Command System failed
3	14 Dec 66	8		Ultra-thin-base film introduced (5,000 ft)
4	24 Feb 67	8	27	
5	26 Apr 67	0	—	Titan failed; no orbit
6	20 Jun 67	10	24	End of development flight program

During this 11-month period, the Defense Intelligence Agency reported that the photography produced by GAMBIT-3 made it possible,

The very success of GAMBIT-3 created a new, but welcome, problem: the routine return of huge quantities of surveillance-quality photography placed a heavy burden on US photointerpretation capabilities. The US Intelligence Board found it necessary to constrain surveillance operations to a manageable level; in January 1967 it scheduled GAMBIT-3 for six FY-67 launchings, with 10 for FY-68, nine for FY-69, and seven for each subsequent year.[77]

In September 1966, Colonel King, who had seen GAMBIT-3 through successful development, was transferred to command the Air Force Satellite Control Facility (he was subsequently promoted to the rank of brigadier general). During the first two GAMBIT-3 flights King was "dual hatted," in that he retained responsibility for the direction of those flights in addition to his duties in the AFSCF. He was replaced as director of the GAMBIT System Program Office by Col.

In August 1966, the NRP Executive Committee decided that, effective with GAMBIT-3 vehicle No. 23, certain configuration improvements should be made. One of the major innovations in what would eventually be called Block-II of GAMBIT-3 was the change from one recovery capsule to two (CORONA had demonstrated the feasibility and utility of using two RVs). In addition, the roll-joint for Block-II would be capable of a minimum of 7,000 position changes, would have redundant mechanisms, and would be able to compensate for the change in PPS inertia which occurred when the first RV was separated. Block-II ultimately included the Air Force's standard space-ground link subsystem (SGLS) for command and telemetry, a completely redundant on-orbit attitude-control system (introduced on vehicle No. 16), and an improved memory in the command processor.

GAMBIT-3 Flight Summary—1967-69

G-3 No.	Launching Date	Photographic Days	Best Resolution	Remarks
7	16 Aug 67	10		
8	19 Sep 67	10		
9	25 Oct 67	10		
10	5 Dec 67	11		
11	18 Jan 68	10		Parachute failed to deploy
12	13 Mar 68	10		2,250-capacity roll-joint installed
13	17 Apr 68	10		
14	5 Jun 68	10		Shortened photographic "burst" times
15	6 Aug 68	10		
16	10 Sep 68	10		Redundant attitude-control system installed. Color photographs taken
17	6 Nov 68	10		
18	4 Dec 68	7		
19	22 Jan 69	10		
20	4 Mar 69	10		
21	15 Apr 69	10		
22	3 Jun 69	10		

Colonel ▓▓▓ retired from the Air Force in June 1968 and was replaced by Col. ▓▓▓▓▓▓▓▓▓, who, while conducting the GAMBIT flight program with a launching about every two months, brought to fruition the important but difficult Block-II improvements. In mid-June 1969, Brig. Gen. William A. King, Jr., returned to SAFSP as its Director, replacing Maj. Gen. John Martin, Jr.

GAMBIT-3 was performing to the satisfaction of its "customers" during this period. In fact, a very close dialogue evolved between the Intelligence Community and the NRO on how to optimize GAMBIT's capabilities. Joint investigations into priority weighting, weather operating thresholds, orbital case development, and optimum launching times were undertaken and closer working relationships resulted in improved satisfaction of intelligence needs.

Occasionally, the Intelligence Community had to be reminded that GAMBIT-3 was specifically a surveillance instrument, unsuited to certain

Dual-Recovery Module

"requirements." In September 1967, for instance, there were suggestions that GAMBIT be assigned the task of collecting mapping, charting, and geodesic information ▓▓▓▓▓▓▓▓ DNRO Flax wisely called for a preliminary study of GAMBIT-3's adaptability to such a task and learned, from his analysts, that 20 dedicated satellites would be needed for the project. The requirement was subsequently revised and ultimately satisfied by other means.[78]

GAMBIT-3 No. 12 contained components which could provide a first step toward developing a double-recovery-vehicle-GAMBIT. It also carried a modified roll-joint with a capacity of ▓▓▓ roll maneuvers (compared to its predecessor's capacity of ▓▓▓).

The "shortened photographic 'burst' times" on vehicle No. 14 were a tribute to the improved accuracy of target location: one no longer needed as much "insurance" footage *around* the targets. Better knowledge of target locations (to exact values) became available because of photographs produced by the CORONA program. Shortening the photographic burst meant that film could be conserved for covering additional targets; from this point forward, the number of targets covered increased significantly (note the results of GAMBIT-3 No. 15, for example).

Exercising Program Priorities

Recovery operations in the early 1970s used C-130 aircraft and range ships. An NRO agreement with the US Navy provided for the Navy to support these operations with two such range ships. As the time for a GAMBIT-3 operation approached, it developed that one ship was in dry dock. The GAMBIT program officers requested Navy support, using their channels to the Office of the Commander-in-Chief Pacific Forces (CINCPAC), which controlled all DoD assets in the Pacific Theater. They were told, in response, that CINCPAC requirements in Southeast Asia precluded providing the usual complete recovery support. Faced with this potential threat to success of the mission, King requested support from DNRO John McLucas, who, in turn, took up the matter with the Chief of Naval Operations (CNO) who, in turn, sent a flash precedence message, in the clear, to CINCPAC. As a result, CINCPAC operations signaled back to SAFSP: "We don't know whom you know, but how many battleships do you want and where do you want them delivered?"[79]

The Block-II Series

All flights in the 1969-72 period were made by Block-II vehicles. In addition, a battery was added and an improved attitude-control system was used. Some difficulties were encountered in the early days of Block-II; a few of them were serious enough to cause loss of imagery. After the successful flight of the first Block-II vehicle (GAMBIT-3 No. 23), a series of malfunctions occurred, all of which were resolved. On vehicle No. 24, the failure of a relay in the vehicle flight control kept the Agena main engine from shutting down on schedule. This resulted in an apogee of 408 nm instead of the desired 220 nm, requiring excessive use of the secondary propulsion system. The AFSCF

Air-Recovery of GAMBIT Capsule by C-130

remote tracking stations had some difficulty locking on to the vehicle transponder; this resulted in delayed command loading and some loss of photography.

GAMBIT-3 Flight Summary—1969-72

G-3 No.	Launching Date	Photographic Days	Best Resolution	Remarks
23	23 Aug 69	15		First dual-recovery capability. First Block-II vehicle
24	24 Oct 69	14		
25	14 Jan 70	14		Second RV failed
26	15 Apr 70	14		
27	25 Jun 70	18		Command System failed; no recovery of second RV
28	18 Aug 70	18		Suez cease-fire zone photography
29	23 Oct 70	18		
30	21 Jan 71	18		First Atmospheric Survivability Test (VAST)
31	22 Apr 71	19		
32	12 Aug 71	22		
33	23 Oct 71	24		High-density acid used in Agena
34	17 Mar 72	24		
35	20 May 72	0		Pneumatic regulator failure during ascent. Total loss. Debris located in England.
36	1 Sep 72	27		Last Block-II vehicle

On the 25th and next flight, the parachute in the second RV failed and the capsule sank before it would be reached by the recovery team. While vehicle No. 26 had no major problems, its photographic quality achieved a best resolution of ▓▓▓▓▓, the poorest in the preceding 18 months.

Problems did persist. On vehicle No. 27, everything appeared normal up to recovery of the first RV. The recovery crew discovered that an ablative shield had failed to separate; however, the parachute and the air-recovery equipment were able to support the extra load successfully. Shortly after commencing the second half of this mission, the heater in the clock circuit of the command programmer malfunctioned. This made it impossible to plan a precise recovery of the second RV. An emergency recovery was attempted but it failed and the second RV was lost. In spite of these disappointments, Colonel ▓▓▓▓▓ and his staff soon gained confidence, as the next seven flights (vehicle Nos. 28 thru 34) performed as expected, with gradually improved photography and increased lifetimes on orbit. These satellites had been scheduled for an orbital mission life of 18 days and they all achieved it (other than vehicle No. 28, which was given orbital adjustments to provide special Middle East coverage and was called down after 16 days, with photography of the Suez cease-fire zone).[80]

Several important personnel transfers occurred during the spring of 1971. Colonel ▓▓▓▓▓ was transferred from his GAMBIT post to become commander of the ▓▓▓▓▓▓▓▓▓▓▓▓▓▓▓▓▓▓▓▓▓ He was replaced at SAFSP

USAF Col Lee ROBERTS

by Col. Lee Roberts, who had been serving in the Satellite Test Center (STC)—a component of the AFSCF—as GAMBIT field test flight director (FTFD). In his FTFD position, Roberts had become throughly familiar with the program, and particularly with problems associated with bringing GAMBIT-3 Block-II on line (in the STC) with new mission control software.

Then, in April 1971, General King was replaced as Director, SAFSP, by Brig. Gen. Lew Allen, Jr., who came to the post from SAFSS, where he had been staff director. Prior to that, he had headed General Martin's Advanced Technology Office (SP-6) in SAFSP. (Allen later became Chief of Staff of the US Air Force.).

Meanwhile, a new development was started at Eastman Kodak for improving photographic resolution.

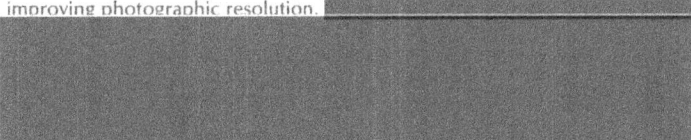

resolution targets were deployed on the ground primarily in desert areas of the southwestern United States. Photographic satellites passing overhead imaged the targets and the film was later used to evaluate on-orbit system performance and to calibrate future payload equipment.

At the end of a GAMBIT-3 mission (with both capsules returned), it was standard procedure to use the Agena's multi-start feature to drive the spent

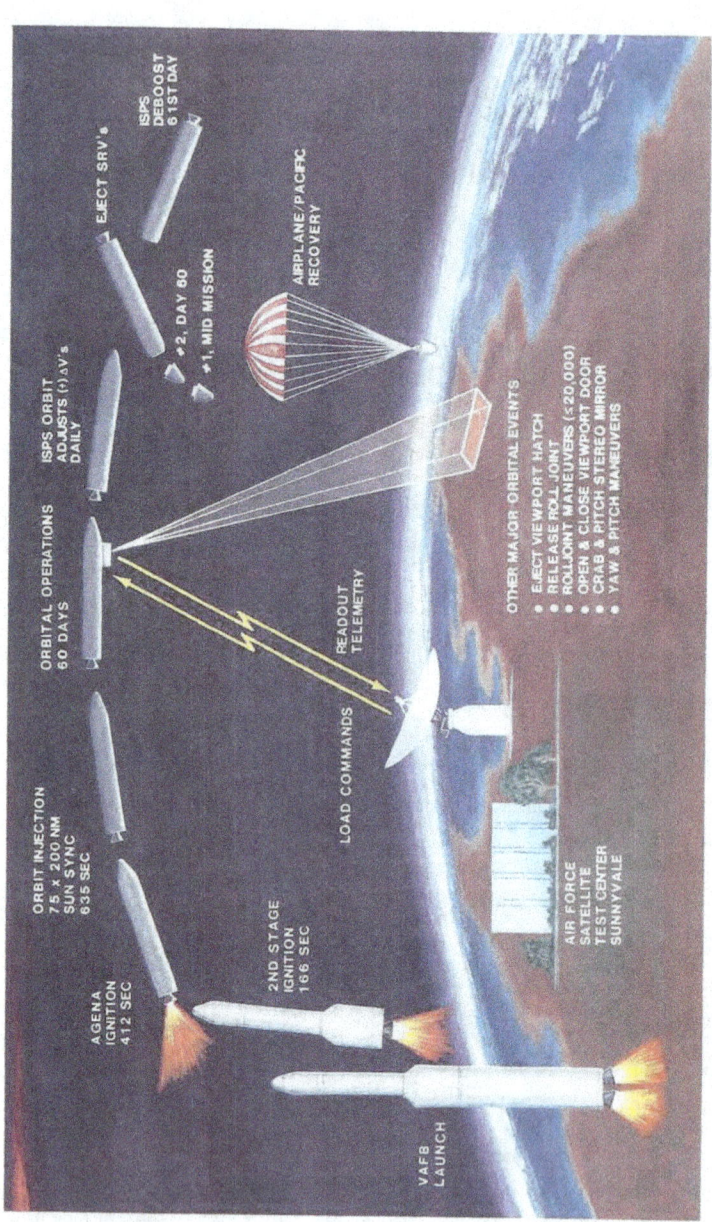

GAMBIT-3 Ascent and Orbital Events

satellite ▬▬▬▬▬▬▬▬▬▬ There had always been a question in the minds of intelligence specialists as to whether surviving bits and pieces existed—debris which might compromise the security of spacecraft design or reveal US technology to other nations. At the conclusion of vehicle No. 30's operation—as well as on vehicle Nos. 31, 32, and 34—atmospheric survivability tests, called VAST, were conducted to learn the extent and nature of such debris. The results of these tests were negative and reassuring, until the next flight—No. 35.

On 20 May 1976, ground stations lost contact with GAMBIT-3 No. 35 (which showed pneumatic-regulator failure) during ascent. As usual, an attempt was made to predict the impact point and a zone over South Africa was indicated. Five months later, Dr. Walter F. Leverton, an Aerospace Corporation employee who had worked on GAMBIT-3, was visiting the London office of his company, where he heard that some "space material" had been found on a farm 75 miles to the north. He arranged a visit to the Royal Aircraft Establishment at Farnborough, where debris was displayed on a laboratory bench. He found three classes of objects: a spherical titanium pressure vessel, some circuit boards of US manufacture, and several chunks of glass which could be arranged into a pie-shape. The glass had the characteristic ▬▬▬▬▬▬ used by Eastman Kodak on GAMBIT optics and ▬▬▬▬▬▬ was convinced he was looking at GAMBIT debris. Eyewitness accounts also strengthened his belief: the objects had been seen falling to earth on 20 May.[83] Discreet arrangements were made for the transfer of these materials to the United States, where they could take their place with the debris from CORONA No. 77—another "errant bird"—which had landed in Venezuela in 1964.

Block-III GAMBIT-3 vehicles had, as their most significant change, a new roll-joint capable of handling 18,000 maneuvers per mission (on GAMBIT-3 No. 12, in 1968, a ▬▬▬ roll capability had been considered "high").

GAMBIT-3 Flight Summary—1972–76

G-3 No.	Launching Date	Photographic Days	Best Resolution	Remarks
37	21 Dec 72	31	▬▬▬	First Block-III vehicles
38	16 May 73	28	▬▬▬	▬▬▬
39	26 Jun 73	0	▬▬▬	Bell engine fuel valve failure
40	27 Sep 73	30	▬▬▬	
41	13 Feb 74	30	▬▬▬	
42	6 Jun 74	45	▬▬▬	
43	14 Aug 74	45	▬▬▬	
44	18 Apr 75	46	▬▬▬	
45	9 Oct 75	50	▬▬▬	
46	22 Mar 76	56	▬▬▬	
47	15 Sep 76	51	▬▬▬	

GAMBIT-3 No. 37, with its 31 photographic days, set a new record for operational time on orbit. ▬▬▬▬▬▬▬▬▬▬▬▬▬▬▬▬▬▬▬▬▬▬

Photograph of Forward Section of GAMBIT Spacecraft

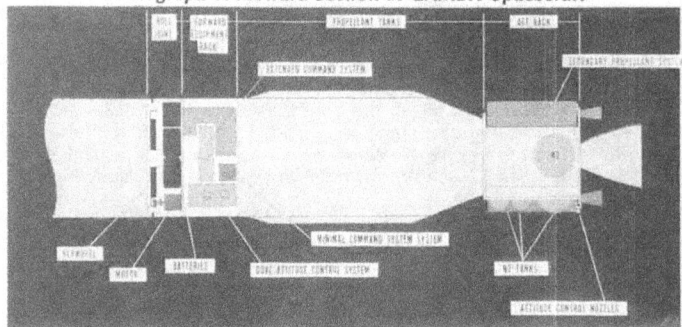

Schematic of GAMBIT Spacecraft

Photograph of Aft Section of GAMBIT Spacecraft

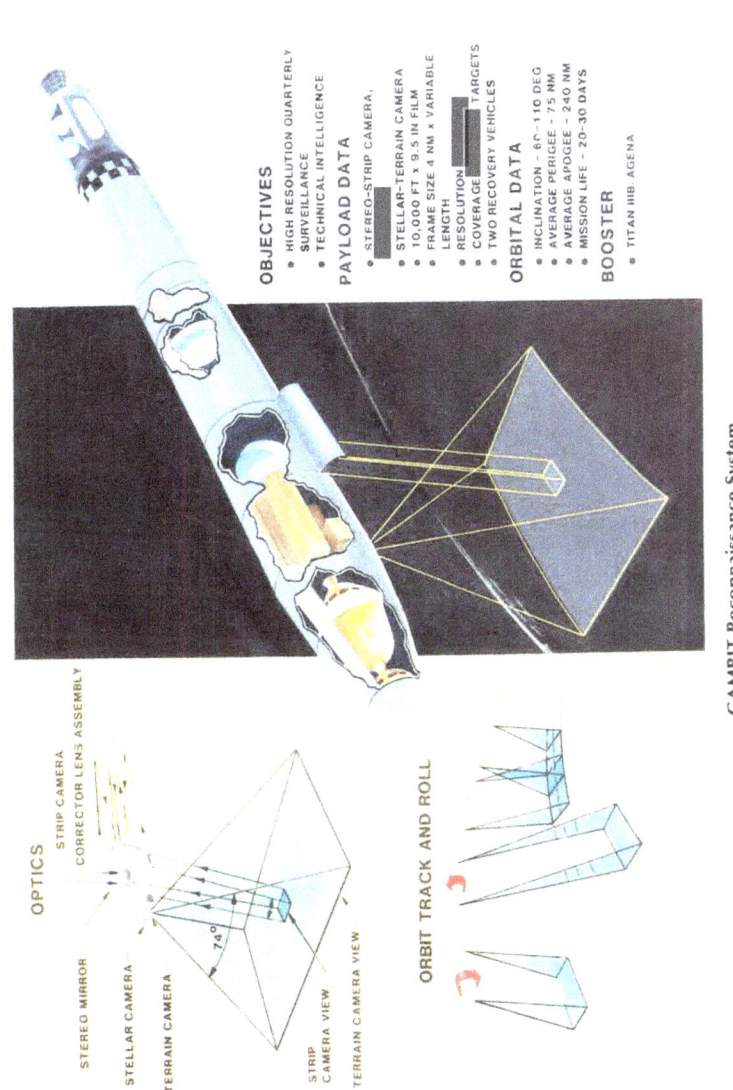

GAMBIT Reconnaissance System

The final block change (Block-IV) was made for vehicles 48 through 54 Major improvements for the PPS were the dual-platen camera, an improved film drive, a new elastomeric heat shield, a new focus system, improved redundancy and reliability, an improved parachute thermal cover, and increased telemetry. Major improvements in the SCS were a main engine baffled injector, solar arrays, added ascent-phase redundancy elements, added cross-strapping capabilities for redundant elements, improved failure-detection circuits, reduced single-point failure potential for the main engine, and the elimination of single-point failure in the roll-joint.

Other than those major design changes to the roll-joint (payload-adapter section) brought about by the addition of the second RV, introduction of the dual-mode mission concept, and increased system redundancy requirements, the remaining modifications were aimed at improving overall system performance. The net result was increased component reliability and greater flexibility in system capability. It was this carefully-planned improvement program that ultimately produced a roll-joint flight system qualified for an operational life of over 20,000 cycles per mission.

The Dual-Platen Camera

At the initiation of the GAMBIT-3 program, Kodak's proposal for the photographic payload had included a camera which had two platens (a platen

NRO Director John L. McLUCAS

USAF BGen David D. BRADBURN

is the device which carries the film behind an exposure slit). At the time, SAFSP Director General Greer considered the dual platen too complex for high confidence in the development schedule and recommended to DNRO McMillan that only a single-platen camera be developed. McMillan agreed. In 1972, the dual-platen concept was revived by both Kodak and Colonel Roberts; its development was started in early 1973, after approval by General Allen and DNRO John L. McLucas, (Dr. McLucas came to the Pentagon from the presidency of Mitre Corporation and replaced Dr. Flax as DNRO in March 1969. While DNRO, he also served as Under Secretary of the Air Force.) The dual-platen camera exposed, independently, both 9-inch and 5-inch film. It provided added versatility to the photographic subsystem by permitting (principally for the 5-inch film) use of other than high-resolution black-and-white film, including experimental films, color film, and false-color infrared. It also provided for adding over 3,800 feet of film for the 5-inch camera. While the use of a riveted assembly was originally planned, problems in the necessary precise alignment of the two platens encouraged change to a welded structure (in some places). The dual-platen camera was first flown in vehicle No. 48, in March 1977, and achieved a best geometric mean resolution

All in all, the photographic performance of GAMBIT-3 continued to meet, or exceed, expectations, with only a few minor problems. One that plagued a number of missions in the 1968–72 era was the effect of minute foreign particles in the exposure slit, causing negative density lines in the negative. This problem was resolved by significantly increasing film-roll cleaning efforts to remove minute film fragments produced when Kodak sliced 56-inch-wide production rolls into operational widths. These fragments were difficult to remove because of their electrostatic charge.

GAMBIT Films

The use of ever-improved photographic films was a positive factor in the evolution of the GAMBIT product. Developed by Kodak (on its own), these films evolved (for black-and-white) from the original Type-3404 film through Type-1414 high-definition film, SO-217 high-definition fine-grain film, and a series of films called "mono-cubic dispersed" or "monodispersed," on which the silver-halide crystals were very uniform in size and shape, providing significantly improved film speed and resolution. SO-315 contained silver-halide crystals of the order of 1,550 angstrom units in dimension; in SO-312 the size was reduced to 1,200, and in SO-409 to 900 angstrom units. In addition to these ever-improving black-and-white films, Kodak produced a line of color and other special films which were flown in lesser quantities than black-and-white, either for special targets or for experimental purposes. Included were SO-121 color film, SO-255 color film, and SO-130 false color infrared film.[85]

The makeup of the actual film load (how many feet, of what emulsion, loaded where in the roll) was determined by the Intelligence Community in conjunction with the selection of mission-target sequences. As the targeting situation was usually very dynamic, it was common for the program office to be "down to the wire" in advising Kodak of the desired composition (in terms of the various films) of the flight roll of film. With the advent of the dual-platen camera the problem of film selection and sequencing became even more complex.

Film-Read-Out GAMBIT (FROG)

From the inception of the reconnaissance-satellite development effort, read-out concepts for imaging systems were considered in the hope of providing new systems with a very quick response time. In fact, this capability had been considered of sufficient importance that three different read-out approaches (E-1, E-2, E-3) were incorporated in the burgeoning Samos program during the late 1950s. Even though the E-1 development finally demonstrated feasibility of the film-read-out concept,* the rather disappointing results (from one successful flight) in terms of resolution (100-foot) and data-transmission rates were certainly a factor in the decision to cancel further development effort. This experience, coupled with the lack of a strong, definitive statement of requirement from the Intelligence Community, discouraged further serious consideration of developing a GAMBIT read-out system until the late 1960s (well into the GAMBIT-3 program). Nevertheless, interest in the read-out idea was kept alive through low-level R&D efforts, which produced steady advancement of the state-of-the-art in key technological areas.

As world tensions mounted during the 1960s, there were periods (such as during the Cuban Missile Crisis, the Soviet invasion of Czechoslovakia, and the Egyptian-Israeli Six-Day War) which stimulated new concern regarding ability to respond adequately to a crisis, which, in turn, revived the question, "Should we have read-out systems?" But once the crisis passed and anxieties quieted, the interest soon disappeared. Typical of this mercurial state-of-mind, a January 1966 COMOR evaluation of the need for quick-response satellite imagery concluded that "the development of a read-out mode for GAMBIT-3 would be worthwhile,"[86] but was followed in few months by a USIB position paper that considered crisis reconnaissance not an urgent requirement.[87] In 1968 and early 1969, however, Dr. Edwin Land, a long-time Presidential adviser on photographic reconnaissance, began to press for near-real-time read-out of reconnaissance photography, supporting the concept of electro-optical imaging (EOI)—in which the system focuses "images" upon a focal plane that directly converts these images into electric signals—as deserving more study.**

---

* The E-1/E-2 systems used the bi-mat technique of processing exposed film by pressing it against a series of web sections containing developer and fixing chemicals. The readout subsystem consisted of a revolving drum associated with a line-scan lens system, a photo-multiplier tube, and a video signal amplifier. The exposed film, once developed on orbit, was optically scanned and the resultant modulated light beam converted into electronic analog signals. Amplified, these signals were then transmitted to the ground station where the process was, in essence, reversed and the image reconstructed.

In May 1969, following discussions with Land, Deputy Defense Secretary David Packard directed DNRO McLucas and the NRP Executive Committee to give serious consideration to read-out system development.[88] The principal sponsors of such work were the CIA's Program B and SAFSP's Program A. Beginning in late 1963, a CIA-funded program known as ▓▓▓ directed by Leslie Dirks, sponsored study efforts to develop an electro-optical imaging (EOI) system. By 1968, ▓▓▓ had been renamed ▓▓▓ and, as a result of Dirks' determined efforts, work was well underway toward perfecting EOI technology. During this same period, SAFSP's Program A was working on Film-Read-Out-GAMBIT, a concept known by its acronym—FROG.* Other possibilities were a less capable, but smaller, device called ▓▓▓ that had been studied by various people, and an electrostatic-tape camera that had never been brought beyond the research stage at the CBS Laboratories.

Deputy Defense Secretary David
PACKARD

▓▓▓ Director Leslie
DIRKS

During the same period, the Intelligence Community, well aware of its inability to react quickly to crisis situations with near-real-time intelligence data, began a preliminary review of the subject under auspices of the Committee on Imagery Requirements and Exploitation (COMIREX), which had evolved from COMOR in July 1967. This review culminated in a 5 January 1968 report[89] which stated, in part: "Our requirements should be interpreted as

---

*Film-Read-Out GAMBIT was essentially the same process used in the E-1/E-2 systems. It used the bi-mat technique of processing exposed film. Once developed, the images would be scanned by a laser device and the resulting data stream would be transmitted to a ground station where it would be recreated by a process similar to that used by the EOI system.

calling for a flexible system that can carry out the warning/indications role and at the same time possess a capability to assist in satisfying routine, current intelligence, and special reconnaissance tasks."

COMIREX also concluded (correctly) that it was the responsibility of the NRO to determine the feasibility of performing a warning/indications mission, from the standpoint of the current and projected state-of-the-art of critical technologies, assessing cost and schedule implications.

After a lengthy evaluation of the various methods of read-out, DNRO Flax, in a March 1969 report,[90] concluded that several promising technical concepts were available, but urged caution in fully embracing those which called for considerable advances in the state-of-the-art, such as electro-optical imaging (EOI). In effect, he reiterated a previous position that "if it were deemed imperative to go for expedited development of a read-out system at this time it would have to be film read-out."

In the summer of 1969, the read-out decision process was complicated further by uncertainties relating to the US position on, and the outcome of, the Strategic Arms Limitations Talks (SALT); it was not clear which might be of more importance to the verification process—higher resolution (better than GAMBIT) or timeliness (read-out). As an additional complication, the NRO was sponsoring some high-resolution work at EK, based upon ▓▓▓▓▓▓▓▓▓▓▓▓▓▓▓▓▓▓▓▓▓▓▓▓▓▓▓▓ now cancelled); this work competed with read-out efforts for NRO funds. It was against this setting that the USIB, on 29 July 1969, approved the requirement for a near-real-time system.

At the 15 August 1969 meeting of the NRP Executive Committee, DNRO McLucas favored a read-out technology development program but recommended delay in choosing a specific read-out approach until the technology would be better in hand. The CIA position, presented by Carl Duckett, CIA's Deputy Director for Science and Technology and Director of the NRO's Program B, was that it was essential to start a read-out program by January 1970, with substantial funds committed immediately to system definition. Duckett's position was supported by DCI Richard Helms. Deputy Defense Secretary Packard proposed a compromise: a more rapid technology and analysis program than that suggested by DNRO McLucas, plus the establishment of a special task force to report to the ExCom on the status of film-read-out technology, electro-optical imaging, and tape-storage systems. The latter item—tape storage—had been studied extensively by Program A as an R&D venture and there was pessimism regarding its readiness for system use. But Dr. John Foster, the DDR&E, wanted tape kept in consideration because, if available, it might prove less expensive and require smaller optics than EOI, and yet give equivalent results. Packard's compromise was accepted.[91]

Dr. Land clearly favored EOI,[92] while the study group set up by Packard—and chaired jointly by Gardner Tucker, the Assistant Secretary of Defense for Systems Analysis, and Dr. Eugene Fubini, a senior advisor to the Secretary of Defense—held that the EOI approach represented a very difficult technology, characterized by the need for very large optics, a large and complex ground station complement, very-wide-bandwidth data-relay equipment for which

DDS&T Carl
DUCKETT

DCI Richard
HELMS

Assistant Defense Secretary Gardner
TUCKER

Eugene
FUBINI

components still were unproven, and an integrating skill that would tax available resources. The Tucker-Fubini Committee noted that a ▇▇▇▇ diameter mirror and ▇▇▇▇▇▇ active electronic circuits were basic requirements for Land's EOI system and that the data-link requirement encompassed ▇▇▇▇▇▇ — which effectively demanded wholly new transmitters, antennas, and specialized components that certainly had to be classified as "beyond the state-of-the-art." Tucker told Packard that in his judgment EOI was too difficult to attempt, as yet, and that instead of approving a system start, the NRO should invest additionally in research and technology improvement. If immediate or near-term results were desired, film-read-out (FROG) was the only feasible route.[93]

In March 1970, after Packard received the report of the Tucker-Fubini Committee, Land reported to Presidential Science Adviser Lee DuBridge that either feasibility experiments or demonstration trials had validated four principal aspects of EOI technology which had been treated previously as high-risk elements. A ▇▇▇▇ diameter mirror with acceptable surface distortion had been fabricated during the ▇▇▇▇▇▇ program for the ▇▇▇▇ camera and a ▇▇▇▇ mirror with a somewhat poorer surface contour seemed readily achievable. Tests of image reconstruction rates had shown that frames containing ▇▇▇▇ of data could be reassembled within ▇▇▇▇ of the time the data were relayed to a ground station, and data-transmission time appeared to be about ▇▇▇▇▇▇ per frame. Laboratory-scale experiments had indicated that mean-time-between-failure (MTBF) rates for individual sensor chips in the solid-state array would approach two years. The Panel had concluded that electro-mechanical devices similar to those used in long-life ▇▇▇▇ satellites would adequately serve other EOI functions and that system MTBFs should, therefore, approach two years. Finally, although the necessary ▇▇▇▇▇▇▇▇▇▇▇▇▇▇▇▇ driver still had to be classified as a high-risk component—its performance not having been demonstrated—the remainder of the data-relay system had, in Land's judgment, advanced to a low-risk category. Land assured DuBridge that a ▇▇▇▇▇▇▇▇▇▇▇▇▇▇▇▇▇▇▇▇▇▇▇▇ was wholly achievable, and that the antennas constituted "no problem." Given that situation, Land maintained it was entirely feasible to schedule a 1974-75 operational date—"if we get on with the development."[94] The points Land selected for emphasis were those aspects of the Tucker-Fubini report that had reached Packard and DuBridge about three weeks earlier; clearly the two groups differed strongly on the issue of EOI technical risk.

Two months later, shortly before the next scheduled ExCom meeting, Land and his associates advised Dr. DuBridge that although both the FROG and EOI approaches to read-out had "reached the stage of demonstrated feasibility and reasonable maturity," the FROG read-out laser-scan system was so complex and limited in growth potential that FROG should be dropped and EOI should be started through the system development process as quickly as possible.[95]

In his 1970 annual report to the ExCom, which arrived two days before the 17 July 1970 ExCom meeting, DNRO McLucas recommended that "essentially all new system effort be focused in . . . the development of a near-real-time read-out system." He also supported a backup effort for the development of a tape-storage camera and continuance of FROG funding which "would deliberately be directed to low-cost, low-risk, and possibly reduced performance systems to provide an alternative for consideration next year."[96]

In his discussions with the ExCom in July, DNRO McLucas expressed concern about selection of the better read-out approach, saying that the system based on a solid-state array might become too expensive in the future, and noting that the ▓▓▓▓ camera was an example of a system being cancelled because it had proven to be more costly than anticipated. After discussion, the ExCom agreed that the total FY-71 investment in read-out development would be ▓▓▓▓ the target date for first launching of a read-out system should be early 1975, and that DNRO McLucas should decide how funds would be spent. On that basis, on 27 July 1970, McLucas authorized the Director of Program B (CIA) to proceed with Phase-I system definition of its proposed system—▓▓▓▓ ▓▓▓▓—on 1 August 1970 at an FY-71 funding level of ▓▓▓▓ and to use the balance of ▓▓▓▓ for tape-storage camera development and for FROG.[97]

In the meantime, COMIREX Chairman Roland Inlow was overseeing an Intelligence Community study effort being conducted under the direction of ▓▓▓▓ DIA member of the COMIREX staff. This study was a comprehensive analysis of crises which had occurred since World War II and the type and timeliness of information required to deal effectively with such situations.[98] In developing its findings, the group had worked closely with Leslie Dirks and his technical staff in DDS&T's Design and Analysis Division, who were responsible for the ▓▓▓▓ EOI advanced technology program. When preliminary

COMIREX Chairman Roland
INLOW

Presidential Science Adviser Lee
DuBRIDGE

results of the ▇▇ study were finally briefed in the Washington area, it was understandable that the stated requirements (that is resolution, frames per pass, frames per day, response time, and so forth) for a hypothetical near-real-time system were closely related to those emerging from the ▇▇ Phase-I definition study. This study, representing the first complete statement of needs for near-real-time imagery, was highly regarded in the Intelligence Community, and, coupled with the Land initiatives during the preceding year, strongly influenced the eventual outcome of the EOI versus FROG debate. The principal attraction of FROG remained its proposed cost ▇▇ against ▇▇ and its two-years-earlier availability.[99]

The only action taken on this matter by the NRO ExCom, at its 29 January 1971 meeting, was to approve continuation of ▇▇ and FROG at then-current rates. It was expected that a decision on full system development could be deferred until November 1971.

On 22 April 1971, a letter from George Schultz, Director of the Office of Management and Budget, informed members of the ExCom that President Richard Nixon had expressed strong interest in having a near-real-time imaging capability at an early date. In response, the ExCom, at its 23 April 1971 meeting, voted to acquire FROG as an interim photographic reconnaissance system for crisis reconnaissance and to delay the proposed first ▇▇ launching until early 1976. Despite this (apparent) final decision, the situation remained confused during the next few months, with Land continuing to push for ▇▇ rather than FROG. After several rounds of infighting, including involvement of key members of the Senate[100] and the Secretary of Defense, on 23 September 1971, Dr. Henry Kissinger, the President's National Security Adviser, advised all concerned of President Nixon's decision to undertake the

President Richard M.
NIXON

National Security Adviser Henry
KISSINGER

development of the EOI ▮▮▮▮ System, with a 1976 operational date, "under a realistic funding program." In addition, there should be no further development of the film-read-out-GAMBIT (FROG) system.[101] With this decision, FROG was dead; Lt. Col. Ralph Jacobson, who had managed Program A's FROG project, asked Col. Lee Roberts for some other assignment in the GAMBIT program and soon became Robert's deputy for payload development. ▮▮▮▮▮▮▮▮▮▮▮▮▮▮▮▮▮▮▮▮▮▮▮▮▮▮▮▮ produced the KH-11 system.

The GAMBIT Award Program

Colonel Roberts worked assiduously and imaginatively to motivate his government-contractor team. As part of his effort to highlight program accomplishments, he visited all of his supporting groups to give informational briefings on intelligence results obtained by GAMBIT, including, where security permitted, copies of actual imagery. He also highlighted GAMBIT problem areas, or "goofs," by originating "The Golden Finger Award." In one typical instance, he presented the award to Eastman Kodak. EK had purchased an expensive and powerful vacuum cleaner to clean GAMBIT flight hardware. Subsequently, the cleaner was used in another area of the plant where, to make it function for another purpose, its three-phase electrical wiring was changed. When the cleaner was returned to its original work site, the wiring was not restored to its original condition and, when next used, it blew dirt into GAMBIT flight hardware. Colonel Roberts, with appropriate ceremony, presented a small gold-plated vacuum cleaner to EK to commemorate a classic example of Murphy's Law at work.

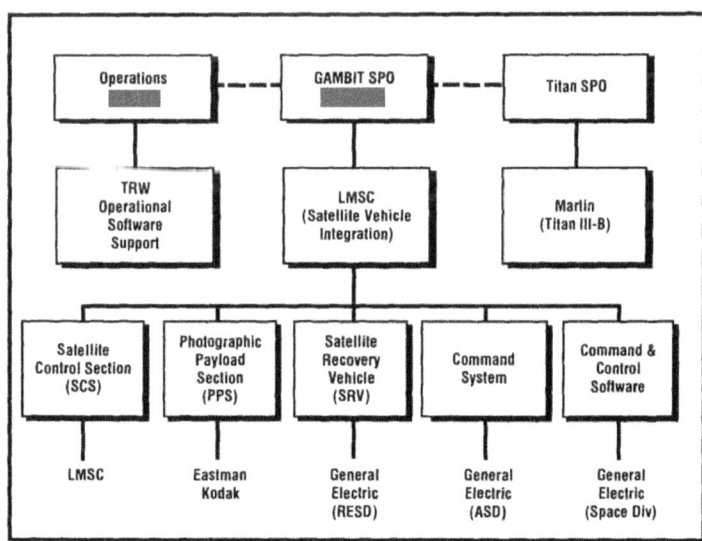

GAMBIT Government-Contractor Relationships

Similarly, after a GAMBIT-3 Agena arrived at VAFB from Lockheed, supposedly flight-ready, but missing two metal nuts in the aft section, Roberts devised another suitable trophy. He sent one of his people to the ship construction area of Los Angeles harbor, to obtain two *very* large steel nuts. After having them covered with gold-colored paint, Roberts ceremoniously presented the nuts to Lockheed GAMBIT Program Manager Bob Koche, who had considerable trouble persuading an airline to accept 150 pounds of extra carry-on luggage.[102]

The NASA 'Gambit'

At the conclusion of the GAMBIT-1 flight program, several "extra" camera systems remained. Appropriately cleared officials at NASA headquarters considered using these in the Lunar Reconnaissance Program (an essential precursor to the newly-assigned Lunar Landing Program). Since the total NASA program was unclassified ("in the white"), it would be necessary to conceal the source and previous purpose of GAMBIT cameras and to midwife them into the "white" world of civilian space flight. In addition, some means of protecting their unique performance characteristics was needed in the (likely) event that lunar photographs were made public. It was hoped that security cover could be provided by the simple expedient of not disclosing flight altitude around the moon, since the lack of such information would frustrate calculation of scale and definition capabilities. Actually, the cameras were never put to use; NASA's own Lunar Orbiter and Lunar Surveyor satellites, already under development, were the only ones to explore the lunar surface. Their resolutions, although quite inferior to GAMBIT values, turned out to be adequate for landing-site selection.

Responding to the Unexpected

The GAMBIT-3 program was not without its "thrills." The success of remedial efforts on GAMBIT flights was due, in large part, to the skill and dedication of the GAMBIT-3 contractor team supporting the program. On the 47th flight, for example, the Titan-IIIB booster, having put the GAMBIT spacecraft on the proper flight vector, failed to back away from the Agena (owing to the failure of Titan's propulsion to terminate properly). Consquently, the aft part of the Agena—its engine programmed to fire—remained in the booster adapter (the interface between the Titan and Agena) with a residual Titan thrust of 1,200 pounds. When the Agena engine fired out of the adapter, the flame destroyed much of the thermal-blanket insulation installed on the Agena's aft rack. The principal purpose of this insulation was to shield propellant lines of the secondary propulsion system (SPS) from the extreme cold of outer space. Without the protection of thermal blankets, the hydrazine in the SPS fuel lines froze solidly by the fourth orbit, resulting in loss of attitude control. Without such control, the GAMBIT spacecraft was unable to maintain its orientation and, within a few days, would have been deorbited by normal atmospheric drag.

Peter Ragusa's Lockheed team assessed the situation quickly and correctly. The vehicle was recaptured in tail-first attitude during orbits six and seven. A reduced mission was accomplished by flying the satellite tail-first, rolled toward the sun to heat the now-exposed plumbing, and then, as needed, reoriented to a nose-first attitude for photography. A test, during the 15th orbit, "indicated that only a maximum of one revolution could be flown nose-first, without the benefit of solar heating of the hydrazine equipment."[105] The need for frequent reorientation of the spacecraft to keep the hydrazine in a fluid state caused a 30-percent reduction in planned photographic coverage. Despite this, the mission ran for 51 days (versus a planned 45 days).

During the 48th GAMBIT flight, two of the spacecraft's five batteries failed; one with sufficient explosive force to cause the vehicle to lose stability on its 45th revolution and to automatically switch over to the redundant attitude-control system. The most probable cause of the failures was "expulsion of zinc-saturated electrolyte solution from battery cells and collection of this solution within the battery case, thus providing a current path to the case. This continuous load raised the battery temperature, causing more electrolyte expulsion so that the failure, once started, is self-perpetuating."[106] The Lockheed support team took steps to reduce vehicle power consumption by:

- Using the redundant attitude-control system starting with revolution 95 to 727;
- Capitalizing on solar-array output, which performed better than predicted; and
- Using standard power-conservation techniques, such as, ▮▮▮▮▮▮▮▮▮▮▮▮▮▮ early in the pass.

The mission was completed without electrical power limitations.

Life-Extension Changes

A number of life-extension changes to the GAMBIT program were undertaken during Colonel Roberts' term as program director. Among these were the addition of batteries and, ultimately, a supplementary solar array which could be extended from the SCS aft rack. To help with the added weight, Colonel Roberts agreed to a change—proposed for all Agenas—to use a high-density acid (HDA) oxidizer to increase the propulsion specific impulse. HDA did work, but caused significant problems because of the effect of the highly corrosive oxidizer on pumps and valves. Even a "simple" change, like adding a solar array to GAMBIT-3 was not without problems, since this required welded solar cells; attaching these to the carrying substrate was surprisingly difficult.[107]

Dual-Mode GAMBIT

In 1966, the HEXAGON/KH-9 program (designed to replace CORONA/KH-4 as a broad-area search system) started its acquisition phase. At the end of the 1960s, concern as to when HEXAGON would be ready and how well it would perform in its flights prompted a number of backup actions. One of these involved the GAMBIT program and was called variously "Highboy," "Higherboy," and, ultimately, "Dual-Mode." Its purpose was to allow the GAMBIT vehicle to fly and operate for 90 days at much higher altitudes—

perigees on the order of 300–350 miles (high mode)—and then to be brought to a lower—more normal—perigee of 78 miles for the balance of the mission. In high mode, the system would be capable of a coverage comparable to HEXAGON. In the Dual-Mode configuration, changes were necessary to both the photographic-payload section (PPS) and the satellite-control section (SCS).

Changes to the PPS included:

- High-altitude photographic capability, including modification of 9-inch and 5-inch frequency-phase-lock-loop electronics to provide slower film drive capabilities; a redesigned film-velocity sensor; modifications to the focus-sensing system; and suitable thermal-paint patterns for both high and low mode;
- Camera automatic-off circuitry and sensor;
- Added smear slits to both 5- and 9-inch cameras;
- Several SRV changes, including increasing total retrorocket impulse by 20 percent, addition of a redundant pyrogen ignition, and increasing the recovery programmer back-up timer interval to 2,808 seconds, allowing high-mode recovery.

Changes to the SCS included:

- Modifications of the main propulsion system (MPS) to provide capability for two additional restarts as a back-up to the integral secondary propulsion system (ISPS), which would nominally be used for all orbit changes;
- Positive ISPS isolation by use of pyro-operated positive-seal isolation valves, in order to maintain the back-up side isolated from contact with the highly corrosive oxidizer until the second side was used;
- Adding a restraint device capable of operating between the roll-joint and the Agena during MPS burns after the initial burns (during the MPS ascent burn the separative joint was held rigid by breakstrips which were separated after injection into orbit);
- Modifications to the attitude-control system (ACS) to change the look-down geometry of the horizon sensor and add a second (high-altitude) commanded pitch-rate.

Dual-Mode was only flown once, on GAMBIT-3 vehicle No. 52. A problem arose: during recovery of the first SRV an electro-explosive-operated flight disconnect failed, precluding recovery of SRV-1. Although the SRV-2 inflight-disconnect-pyro also failed to function, a backup/recovery deboost was effected, using the satellite-control section's ISPS to reenter the entire vehicle, whereupon SRV-2 separated and was recovered. Quality of the imagery from flight No. 52 was degraded nearly 50 percent. Despite many months of investigation by a team from many elements of the program—plus independent outside help—the exact cause of the degraded performance on this flight could not be identified.

GAMBIT Reaches Full Potential

In August 1977, Colonel Roberts was replaced as GAMBIT program director by Colonel ███████████████████ came to the assignment from the ███████████████████ he had also served in SAFSP program activities. By this time GAMBIT, with 48 vehicles flown, could certainly be characterized as a fully mature, successful program. Thus, having realized most of the potential performance to be gained through system upgrading, the Program Office turned to the operational area for additional improvement. Working with the Program A Operations Office ██████ and, through it, with members of the Intelligence Community responsible for targeting requirements (Imagery Collection Requirements Subcommittee [ICRS] of COMIREX), a number of operational refinements were made. With

In January 1983, Colonel ▓▓▓▓ was reassigned ▓▓▓▓ SAFSP, and replaced by Col. ▓▓▓▓ as director of the combined GAMBIT and HEXAGON programs. This combined program management concept had been conceived and initiated during Colonel ▓▓▓▓ tenure as SPO Director. It merged the management, engineering, and test functions of two mature programs, both at SAFSP and at Lockheed Missile and Space Company—the principal contractor—resulting in significant reductions in manpower and building space, which, in turn, translated into substantial dollar savings.

Under Colonel ▓▓▓▓ direction, the last two GAMBIT missions were conducted with distinction. Mission No. 4353, for example, surpassed all previous missions in terms of duration (129 days); number of frames ▓▓▓▓ and gross number of targets ▓▓▓▓. In addition, ▓▓▓▓

GAMBIT-3 Flight Summary—1977–84

G-3 No.	Launching Date	Photographic Days	Best Resolution	Remarks
48	13 Mar 77	69		First dual-platen camera
49	23 Sep 77	73		
50	28 May 78	90		
51	28 Feb 81	110		
52	21 Jan 82	119		Only Dual-Mode mission flown
53	15 Apr 83	126		Longest duration ever—129 days (including 3 days "Solo")
54	17 Apr 84	116		

During these 12 years, GAMBIT-3 improved steadily in time-on-orbit, eventually lasting three to four months on each flight. Resolutions of ▓▓▓▓▓ kept the system preeminent as one of the foremost technical intelligence collectors in the reconnaissance-satellite system inventory. By the time of the last GAMBIT flight, in April 1984, ▓▓▓

Section 6

GAMBIT Program Costs and Highlights

GAMBIT-1/KH-7

The total cost of the 38-flight GAMBIT-1/KH-7 program, covering fiscal years 1963 through 1967, was $651.4 million in 1963 dollars. Of that amount, ▮▮▮▮▮▮▮▮▮▮▮▮▮▮▮ was incurred in SAFSP contracts and the remaining ▮▮▮▮▮▮ in SSD/AFSC and CIA contracts. The $651.4 million includes a ▮▮▮▮▮ cost of hardware purchased for GAMBIT-1 but reallocated by DNRO to other NRO projects; it does not include the cost of five GAMBIT-1 payloads sold to NASA.[111]

"Non-recurring costs for development, industrial facilities, and one-time support totalled ▮▮▮▮▮▮▮▮▮▮▮▮▮ of the total program cost. ▮▮▮▮▮ of the development cost was for development of the satellite vehicle by GE, and ▮▮▮▮▮ for development of the payload by Eastman Kodak."[112]

An analysis of recurring costs only gives average unit costs per GAMBIT-1/KH-7 flight: ▮▮▮▮▮▮▮▮ for the first ten flights and ▮▮▮▮▮▮▮▮ for the last 28. Lumping all costs, both recurring and non-recurring, into a total:

Average cost per flight ▮▮▮▮▮
Average cost per day in orbit ▮▮▮▮▮
Average cost per target photographed ▮▮▮▮▮

It should be noted that cost per target photographed fell to ▮▮▮▮▮▮▮▮▮.

The principal contractors were General Electric for the spacecraft, costing ▮▮▮▮▮▮▮▮▮▮▮▮▮▮. General Dynamics for the Atlas missiles, hardware, and launching expenses, totalling ▮▮▮▮▮▮▮ Lockheed Missiles and Space Company for Agena second-stage vehicle, Agena-peculiars, and launching, amounting to ▮▮▮▮▮▮ and Eastman Kodak for the camera payloads, costing ▮▮▮▮▮. The remaining ▮▮▮▮▮ was for satellite control, command-generation software, mission planning, Aerospace Corporation support, and industrial facilities.

GAMBIT-3/KH-8[113]

The total cost of the 54-flight GAMBIT-3/KH-8 program (fiscal years 1964 through 1985) was $2.3 billion, in respective year dollars.

Non-recurring costs for development, industrial facilities, and so forth, totalled ▮▮▮▮▮▮▮▮▮▮▮▮▮▮ of the total program cost. ▮▮▮▮▮ of the development cost was for the satellite-control section (SCS) and

Agena-related hardware by LMSC, whereas ▓▓▓ was for development of the photographic-payload section (PPS) by EKC and recovery vehicle (RV) by GE. Costs (in millions) for the GAMBIT-3/KH-8 program were:

	Recurring Costs Only	All Costs
Average per flight	▓▓▓	▓▓▓
Average per day in orbit	▓▓▓	▓▓▓
Average per target photographed	▓▓▓	▓▓▓

Principal contractors were Lockheed Missiles and Space Company for the satellite-control system ▓▓▓; Eastman Kodak for the photographic-payload section ▓▓▓; Martin-Marietta for the Titan-IIIB booster and launch, ▓▓▓; General Electric for the command subsystem and reentry vehicles, ▓▓▓.

GAMBIT in Retrospect

From the first GAMBIT-1/KH-7 flights in 1963 to the final GAMBIT-3/KH-8 series in the 1974-84 period, the record of reconnaissance performance was remarkable, certainly meeting President Eisenhower's directive of 1960 that the Air Force should develop a film-return, high-resolution imaging satellite system.

- *Resolution:* Initial results of two to three feet, in GAMBIT-1/KH-7, improved to ▓▓▓ two feet early in the GAMBIT-3/KH-8 phase, and, in the last 10 years of the program resolutions ▓▓▓ were consistently achieved.[114]
- *Coverage:* The number of targets covered by the early GAMBIT-1/KH-7 missions amounted to ▓▓▓ per mission, reaching ▓▓▓ by the 18th flight; the remaining GAMBIT-1 missions covered between ▓▓▓ and ▓▓▓ targets each. Early GAMBIT-3/KH-8 flights acquired over ▓▓▓ photographic frames (often containing more than one target). By the 23rd GAMBIT-3 flight, the number of photographic frames exceeded ▓▓▓ and, by the 41st flight, was more than ▓▓▓. The penultimate GAMBIT-3 flight, No. 53, acquired ▓▓▓ frames, which contained ▓▓▓ targets. By way of comparison, all 38 of the GAMBIT-1 missions photographed only ▓▓▓ targets.[115]
- *Duration:* Early GAMBIT-1 flights flew one- to two-day missions, gradually improving to six and eight days. Early GAMBIT-3 flights were of the order of one week in duration; flight Nos. 6 to 22 generally lasted 10 days. Duration increased, by flight No. 27, to 18 days and by flight No. 32 to 22 days. Flight Nos. 36 through 41 had durations of 27 to 31 days; the number increased to 45 days during flight No. 42 and to 69 days by flight No. 48. The last four flights were in the four-month class.

The 54 GAMBIT-3/KH-8 flights achieved a reliability of over 94 percent; only three failed to reach orbit (two Agenas and one Titan failed). Once the GAMBIT-3 spacecraft got on orbit, it never failed to perform, in spite of a "few" problems. The significant contributor to this remarkable record was the management environment created for GAMBIT. This feature is discussed in detail in the following Section.

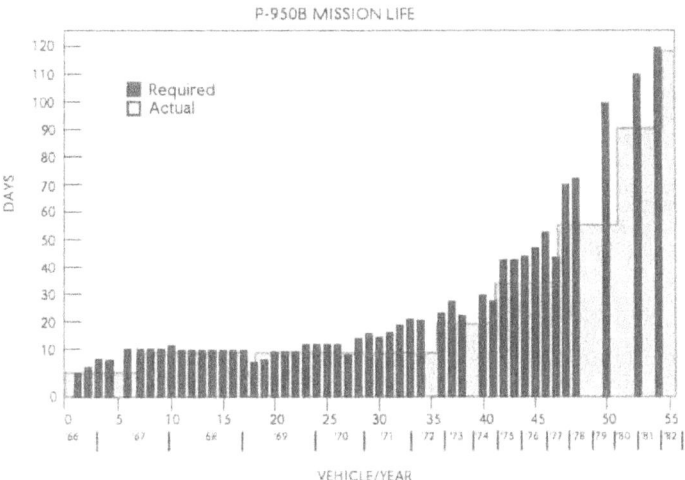

GAMBIT-3 Mission Life Growth

GAMBIT-3 Resolution Improvement

Section 7

Development Management: Styles in Program Control

Without question, the most important US military weapon systems developed during the 1955–65 decade were intercontinental missiles and reconnaissance satellites. Although these systems were produced by Air Force managers, using Air Force facilities, none of them was developed in any of the already existing Air Force research and development centers! There are reasons for this fact—reasons which are a somewhat mordant comment on the status of military development program management.

After establishment as a separate service in 1947, the US Air Force acquired a plethora of technological installations: Wright Air Development Center (ADC) in Ohio dedicated to all elements of aircraft and missile work; Rome ADC in New York for electronic studies and gear; Cambridge Research Center in Massachusetts for research in electronics and geophysics; Holloman Air Force Base (AFB) in New Mexico for missile testing and human factors studies; Eglin AFB in Florida for proving ground tests; and Edwards AFB in California for aircraft testing. It soon added Tullahoma AFB in Tennessee for wind tunnel work; a Special Weapons Center at Kirtland AFB in New Mexico to support nuclear testing; and a missile-testing range at Patrick AFB in Florida.

All of these organizations enjoyed growth beyond anything anticipated in the early postwar years. Places like Holloman, Cambridge, and Rome, which had originally existed as small "stations" subvervient to Wright Field, suddenly became full-scale ADCs, laying claim to technological preserves of their own and lobbying for manpower and dollars to support the claims.

On-Time Delivery: A Look at the Record

It is a truism that "bigger" does not necessarily lead to "better." The growing centers of the new Air Research and Development Command (ARDC), established in 1951, exemplified this saying. The facilities of ARDC's centers were improved, the rosters of projects assigned and undertaken grew to impressive lengths, more people were hired, and more dollars were spent; however, the capability of the Command to deliver new or improved hardware to the operational Air Force *on schedule* remained rather stagnant. On the Command's seventh birthday, in 1958, it could cite only one case where a system *had* been delivered on its original schedule: the anomaly had occurred in 1955, at the Cambridge Research Center. Hq, ARDC, then under the salutary command of Maj. Gen. Thomas S. Power, had inquired into the reasons for this unusual event:

> It would be of value to this Headquarters to receive from the ARDC member of the WSPO [Weapon System Project Office] an informally written description of those operating principles which his experience has indicated contribute to the successful and on-time production of a system. To be most useful, the comments should be completely frank and candid.[116]

SECRET
Handle via
BYEMAN-TALENT-KEYHOLE
Control Systems Jointly
BYE 140002-90

It was ironic that the addressee, Cambridge, had only one assigned system, being largely what its name implied: a research center. On the basis of a perfect batting average, the response to the request must have been a pleasure.

The major player, by far, in *weapon system* development, was Wright Field and practically all of ARDC's system development was assigned there. Consequently, Wright Field had the major concentration of ARDC manpower and money. WADC managers, knowing that they had a serious problem with on-time delivery, had sponsored several studies in searching for possible solutions. The Belden Study[117] (previously mentioned in this volume) was the best known of these. WADC's Thomas G. Belden had carefully examined the history of 100 key developments under WADC's aegis and had found that

- 85 slipped 0.45 year or more per year, and
- 22 (of that 85) slipped one year or more per year.

Air Force "customers"—the operating commands—learned of this analysis and were understandably disturbed by its implications. Strategic Air Command planners, for example, could see that if they placed an "order" with WADC for development-delivery of an item of new hardware, and it happened to fall within the first of the Belden-categories, an originally agreed-to delivery forecast of, say, four years, could slip, on the average, to seven-and-a-half years, eight years, or longer. If the order fell into the second category, SAC could assume, in advance, that it would never be delivered!

The WADC situation was not unique. Throughout ARDC, "in-spec, below-cost, on-time" delivery appeared to be a coveted but unattainable goal. When challenged on the matter, Centers became irritated and defensive, usually countering with references to people and dollar-shortages ("If we were *just* given the tools to do the job," and so forth). But the ARDC's annual budget and population figures *were* actually increasing each year; sadly, the delivery slippages themselves devoured much of the Command's resources, and their costs were increasing annually.

The On-Time Delivery Problem: Contributing Factors

It was a sad fact that ARDC Centers had little experience in, little motivation toward, and little inclination for, working in what industry called "a short-leashed environment"—one where missions and careers would rise or fall on the basis of production performance. Center concerns and motivation lay elsewhere: the primary goal was organizational (and, therefore, personal) stability and security.

Center staffs and laboratory people were preoccupied with establishing and policing territorial franchises for very broad areas of technical activity. This reflected in program budgets, where many tasks never actually ended and additional work was continually proposed. Residual time was spent in "monitoring" contractors' work results—essentially reading reports—and examining expenditures (in-house technical laboratory work usually proceeded with a minimum of supervision). Center-staff members, one hierarchic step above laboratory workers, did not monitor—they "maintained cognizance." Further up, at the summit, the leadership of the Center "exercised broad staff surveillance."[118] Responsibility for meeting deadlines was usually shared, blurred, or unassigned—the classic bureaucratic stance. In this environment, a Center's activities culminate inevitably in a program which, by commercial contractor standards, would be classed as leisurely, undisciplined, and expensive.

Although any Center could have an on-time delivery problem, only Wright Field was in the *system* development business to a major extent. The delivery problem, like other problems, was caused by people—both military and civilian. The bulk of Center manpower was, of course, civilian and the rules composed by the Civil Service Commission to govern the hiring, promotion, and firing of civil servants applied to the Center. As years passed, these rules seemed to augment their (natural) bias in favor of the employee. Indolence in matters technological or *sangfroid* regarding failure in on-time delivery were not listed as grounds for replacement or dismissal of a civilian. In fact, acceptable grounds for disciplinary action, on any basis, grew more and more narrow, with heavy burden of proof on the supervisor. Bringing such actions required the "accuser" to show a detailed journal, often for as much as two or three years, cataloging explicit deficiencies, citing extensive counseling efforts, and actually *proving* that the employee had shown very little or no positive response. The most persistent and exhaustive disciplinary efforts, at Center levels, would be followed by long-drawn-out reviews at Command and Headquarters, Air Force levels, where reversals were commonplace. And, since the action had to pass initially through the Center's headquarters, which was staffed mainly by military people, the process triggered considerable schism between the military and civilian "households."

As years passed, and examples of disciplinary *action* failed to accumulate, there developed a grapevine consensus that "getting rid of" a civilian was essentially "too hard." In addition, it was potentially career-threatening to the Center commander and staff. Inevitably, strong disciplinary action became a rarity in ARDC and its Centers were considered havens of "locked-in" security for civilian employees. There developed a world in which there could be rewards, but very few punishments; a world in which promotion tended to be based on how many persons one supervised, rather than on how many "on-time" deliveries he had made.

The other group of Center employees—the military—brought their own unique problems. Primary among these was the brief tenure of research and development officer-specialists. Even the best-trained officer would need a least a year to learn the technical program to which he was assigned. Then he would probably have, at most, two remaining years to contribute to the work at hand. During his "productive" period, he would have frequent outside demands: special military assignments (courts and boards), military training seminars, flying-time requirements, and assistance to the commander's staff in many base activities. Then he would be transferred out of the Center (the civilians could say, "Just as he became useful to the program") to a service school, or a career-enhancing next assignment. All in all, the military engineer was a limited asset, as fleeting and evanescent in his impact on the Center as the civilian's was fixed and stolid.

AFBMD: High-Level Response No. 1

This picture of the Air Force's research and development household in the 1950s was fairly well understood at the highest governmental levels, largely because all really "big" on-time delivery problems automatically come to the attention of top Air Force and Defense officials. With the introduction of a Presidential Science Adviser, during the second Eisenhower administration, this kind of bad news was also known, rather swiftly, at the White House.

In 1954 and 1955, when the requirement for on-time delivery of intercontinental and intermediate-range ballistic missiles became a primary governmental urgency, it was agreed in Washington offices that routine assignment of such a task to an existing ARDC Center would not even be discussed. So a new organization—the Western Development Division (WDD)—was created to undertake the responsibility. WDD was to be part of the Headquarters, ARDC commander's office and would by-pass the Air Staff, in its command line, reporting directly to an Air Force Ballistic Missile Committee (chaired by the Secretary of the Air Force) and, subsequently, to an Office of the Secretary of Defense Ballistic Missile Committee. Bernard A. Schriever, then a brigadier general, was appointed to command the new division. Schriever had sufficient experience with "normal" research and development processes to press for the privilege of requesting his officer-staff by name, to insist that he control their tenure, and to state, very discreetly, that he would probably not appoint civilians to his staff.[119] It also became evident that he did not intend to draw upon existing ARDC Centers for technical support; rather, he would hire a contractor (Ramo-Wooldridge) to furnish essential technical direction and system engineering. He would, of necessity, use ARDC's test centers—particularly the AF Missile Test Center (and range) at Patrick AFB.

CORONA: A Second High-Level Response

When the need for a reconnaissance satellite became an acute emergency, President Eisenhower, in February 1958, placed the CORONA program jointly within the Air Force Ballistic Missile Division (AFBMD), the new name for the WDD, and the Central Intelligence Agency, permitting no intermediate review or approval channels between the project and himself. In this arrangement, one person each, from the Air Force and the CIA, would report directly to him. This revolutionary management arrangement was the ultimate in establishing direct lines of authority and responsibility for on-time delivery of a major system.

AFBMD's responsibility for engineering and operating the CORONA system was "covered"—in a security sense—by its organizational designator: "Discoverer Office." Lt. Col. Lee Battle, a specialist in propulsion engineering, headed this office. His professional experience had been acquired in ARDC, where he had been a keen observer of management styles and processes and where he had privately developed his own detailed "Battle-model" for how things should be done. In his Discoverer-CORONA post, he saw an opportunity to finally put his ideas into effect on a major scale and he welcomed the occasion. Battle listed his management principles (in order of importance) and his subordinates soon knew them by heart.[120]

1. *Keep the program office small and quick-reacting.* The Discoverer Office, in 1958-59, consisted of Battle plus three others; in 1960, with the program in full operation, the total manpower was five. Each individual had absolute responsibility for an explicitly defined area. Battle insisted on people who could be "short-leashed," energized, and, if necessary, replaced in 30 seconds—in other words, he wanted military people.
2. *Select your people with great care and then rely on them.* Fortunately, the AFBMD's work had a high priority which made such personnel selections possible. "Relying" meant assigning entire responsibility for a technical area to one person and then holding him (not the contractor) primarily responsible for success.
3. *Control the contractor by direct personal contact.* The responsible officer was to know what the contractor was doing, in his assigned area, *every day.* He was to assess the contractor's key people and their work continually, remembering that personnel errors are much more frequent than design errors.
4. *Make as much use as you can of (external) supporting organizations.* This was a device for keeping one's own staff small. (Battle would add, *sotto voce,* "To insure adequate support always make unreasonable demands.")
5. *"Hit" hard on checkout and flight failures.* "Unfixed" problems will rise to bite you again. Reject the expression "random failure"; there is no such thing.
6. *Cut out as much paperwork as you can.* Comply promptly with mandatory reporting requirements—*in the most meager fashion that will be accepted.*
7. *Do not over-communicate with higher echelons.*
8. *Avoid committees.* There is always an individual to whom you should have given responsibility for what the committee thinks it is doing.
9. *Rely on your contractor's technical recommendations, once you are sure he has given the problem sufficient study.*
10. *Have very close ties between your office and your representatives in the field (at the launching site or satellite control center, for example).*

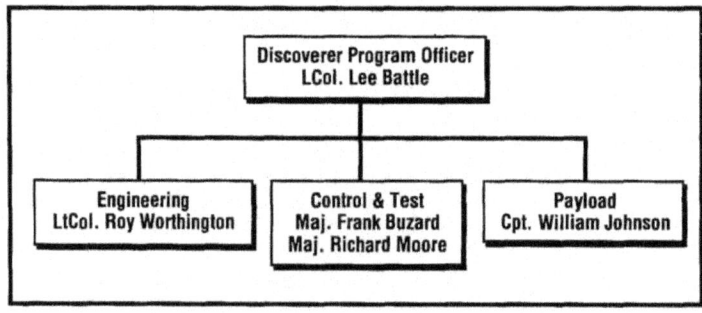

The Discoverer Program Office, Spring 1960

The success of Battle's spartan group was enhanced, of course, by some outside factors. His work was recognized at the highest governmental levels as absolutely essential to the national welfare; therefore, he had strong support from above: a top priority and assured funding. He could name-request and "freeze" his engineers; he could get instant reactions from supporting organizations; and he could insist on contractor-colleagues who shared his management views and sense of urgency. When the program hesitated or faltered, in the early flight series, his great and good CIA friend Richard Bissell served as proxy in Washington, D.C.—even for "woodshedding" at the White House.

SAFSP: The Third High-Level Response

The reasons for establishing a Secretary of the Air Force Special Projects Office (SAFSP) in Los Angeles and, finally a National Reconnaissance Office, with a Program A at SAFSP, have been covered in this volume. Significantly, while it had seemed proper and feasible to place CORONA development in the context of the (relatively new) AFBMD organization, when the time came to accelerate satellite reconnaissance *follow-on* systems, a new organization was created. GAMBIT was the first full-scale venture of the National Reconnaissance Office, and, while it was to proceed independently of CIA "godfathering," it would enjoy some similarities: an equally high national priority, the license to name-request and retain its people, and a determination to achieve *on-time* successful operation. The SAFSP part of the NRO was staffed with Air Force officers, most of whom had professional backgrounds very similar to Battle's and some of whom were aware of, and very much impressed by, the CORONA record of achievement. SAFSP, and its GAMBIT program office, acknowledged a debt to CORONA as pathfinder and proposed to do as well, or better, in its own developments.

In addition, Generals Greer and Martin, together with Colonel King, were veterans of the Air Force research and development community and deeply interested in achieving the best possible management style and procedures. The principles SAFSP followed reflect much of Battle's philosophy, with several extensions of theme.

1. *Keep the program office small*—not in order to save manpower but in order to encourage strongly personal interactions.
2. *Hand-pick your people;* stick to known military "winners," use name-requests and control tenure.
3. *Control contractors by direct personal contact,* rather than by paperwork. Place top-notch officer-engineers at the key contractor plants—particularly those where technical problems are the most severe (GE Philadelphia; Eastman Kodak). The officer's function is *not* the classical procurement role: checking contract compliance. Rather, he is to (a) know exactly what the SAFSP GAMBIT Program Office wants and needs, (b) what the contractor plans to do, or is doing, to satisfy those needs, and (c) develop such a strong rapport with the actual "doers" at the plant that they will voluntarily tell him their problems and proposed "fixes," in advance, or at the latest, as soon as they become known, and (d) stay in *daily* communication with the GAMBIT (home) office. (This

guidance was followed so successfully that King's representatives were usually invited to most contractors' meetings.)
4. *Stress that program success is the raison d'etre for the SAFSP and GAMBIT Program Office.* Leadership in stressing this fact was furnished by Generals Greer and Martin as they built a tradition of personally and faithfully participating in every GAMBIT flight operation: at Vandenberg AFB (launching) and at the Sunnyvale Satellite Control Facility (on-orbit operations). In establishing this "presence," the generals were very careful to keep the atmosphere familial, rather than hierarchical. They wanted it understood that they were present because successful flight operation was the sole purpose, and a culmination, of their organization. They also wanted it understood that the Program Director, for example, Colonel King, was in *complete charge.* If he wished to have consultation on a flight problem, the generals were available to help him ponder. But they were not there to take command, or to cast a shadow on the operation. The process worked superbly; the generals gauged its success by the fact that the contractors scarcely seemed aware of their presence.
5. *"Keep It Simple."* As frequently illustrated by example in this volume, "Keep It Simple" was Colonel King's by-line, extending to everything he directed. He exemplified this by:
   a. Using proven components whenever possible (CORONA's capsule, Lifeboat, the roll-joint).
   b. Trimming non-essential engineering.
   c. Buying fewer spares.
   d. Sticking to a single check-out.
   e. Abbreviating documentation.
   f. Simplifying tests.

Applying this maxim strictly allowed King to return ▓▓▓ in FY-63 and ▓▓▓ in FY-64.

A Summing-Up

The ballistic missile and satellite reconnaissance development experiences show strong parallels in (a) what must be *avoided* and (b) what must be *done,* when on-time delivery is a primary factor. There was a remarkable similarity in the basic management style of the programs and positive evolution in that style as improvements were added with each program iteration. It was exhilarating to prove that "in-spec, below cost, on-time" delivery *is* possible in an Air Force environment.

But there was also a sobering element in the knowledge that, just as ballistic missiles could not be produced—*on-time*—in the WADC environment, satellite reconnaissance systems could not be assigned to the AFBMD. Absent constant, stern watchfulness and periodic "purging," it appears inescapable that all governmental research and development organizations eventually follow a well-worn path toward bigness, security, inertia, and incompetence. This unpleasant knowledge has been expressed in three maxims, generally acknowledged and rarely heeded:

- Once an organization achieves a certain—apparently rather modest—size, it can never again have enough people to carry out its mission.
- Once the members of an organization achieve the level of "tenured" security that they think they desire, they can never again carry out their mission.
- Every time a truly urgent technological achievement is required by our nation, a new organization must be assembled.

Section 8

From 'Haunting Concern' to Informed Response

In recalling the busy days of 1955, Dwight D. Eisenhower would surely have given primacy to two events. The first was international in character: the "Open Skies" proposal which he had made to the Soviets at Geneva on 21 July, urging mutual consent for aircraft overflight of each other's territories as a means of allaying "the fears and angers of surprise attack."[121] He had been disappointed and distressed when his offering was rejected—in an off-hand and casual manner—and had summarized the event in a trenchant comment: "Khrushchev's own purpose was evident—*at all costs to keep the USSR a closed society.*"[122]

The other focus of his recollection would be personal in nature (if anything in a President's term is truly personal): in September of that year he had experienced a very serious heart attack. Fortunately, his recovery had been so swift and complete that he was able to run for a second term of office. He enjoyed five years of peaceful retirement before the cardiac problem recurred, in 1965. Once again, his recovery was swift and reassuring. The third attack, in April 1968, was a different matter: he entered Walter Reed Army Hospital at once, staying there until the end of his life, in 1969.

One February day, in 1969, Eisenhower mused to a friend that he missed the excellent intelligence briefings which he had received during his White House days. He went on to wonder, in particular, what improvements might have occurred in the technology of overflight photographic reconnaissance. His friend promised to arrange a "state-of-the-art" briefing.[123] His request was relayed across the city to the National Photographic Interpretation Center, which was presided over by Arthur C. Lundahl. Characteristically, Lundahl welcomed the request; he had briefed Eisenhower several times previously and found him an excellent audience.

Lundahl was respected and honored in the US Intelligence Community as the nation's most knowledgeable and articulate briefer. He was a superb photo interpreter in his own right, and combined his technical skill with a warm enthusiasm for the subject and strong empathy with the audience. He had recognized, early in his career, that the usual audiences—whether military or civilian—should not be expected to have a photointerpreter's insight into what was on the briefing boards, so he had become a master at tailoring presentations to that human condition, helping lay audiences transcend their inherent limitations. Lundahl knew, for instance, that a professional who had studied the USSR's Tyuratam missile launching facility in detail for five years could exclaim enthusiastically over a minor, new construction element, while, to a layman, the object might appear, at most, as a vague blob or blur. The ubiquitous character of this problem was well-illustrated by Robert F. Kennedy's remarks on a momentous briefing, given at the White House on 16 October 1962:

> At 11:45 that same morning, in the Cabinet Room, a formal presentation was made by the Central Intelligence Agency to a number of high officials of the

government. Photographs were shown to us. Experts arrived with their charts and their pointers and told us that if we looked carefully, we could see there was a missile base being constructed in a field near San Cristobal, Cuba. I, for one, had to take their word for it. I examined the pictures carefully, and what I saw appeared to be no more than the clearing of a field for a farm or the basement of a house. I was relieved to hear later that this was the same reaction of virtually everyone at the meeting, including President Kennedy. Even a few days later, when more work had taken place on the site, he remarked that it looked like a football field.[124]

Now it had been absolutely clear to *Lundahl* that the Soviets had, at this location and as of this date, introduced intermediate-range ballistic missiles into Cuba; to his audience, none of whom rose to dispute his analysis, the San Cristobal missile site, as seen from the U-2, looked like a fuzzy farm, a basement, or vaguely like a "football field."[125]

There was also another version of the audience's response, reported to Lundahl later that day, which he (understandably) cherished. This version illustrates his point about photo interpreters and laymen:

RFK to JFK:	Did *you* see the missiles site?
JFK:	Did *you*?
RFK:	Frankly, no!
JFK:	Neither did I, but he was certainly convincing, wasn't he?

Lundahl ranked Eisenhower as one of his best audiences. He knew from experience that the President would follow a briefer's words intently; he had continued the habit, from White House days, of closing in on the photography, from time to time, with a huge magnifying glass and a firm, "Now show me exactly *where* that is and *why* you call it what you did."[126]

Lundahl selected photographs for the Walter Reed briefing, using two criteria: he wanted to update Eisenhower on the most important developments at denied foreign locations and he wanted to vignette technological improvement in the intelligence enterprise which Eisenhower had boldly godfathered during his White House years. Lundahl had no qualms about including U-2 photography in his briefing; he wanted to reassure Eisenhower that—in spite of one distasteful episode—the U-2 had been, and continued to be, a primary intelligence collector. Its more sophisticated offspring—the A-12 OXCART (predecessor of the SR-71)—had also overflown vast areas, even monitoring the captive USS Pueblo in North Korea's Wonson Harbor. He chose photos obtained by the Ryan-147 drone to show that not all collection had to be done at high altitudes, particularly when one was supporting ground forces. Finally, he selected CORONA/KH-4, GAMBIT-1/KH-7, and GAMBIT-3/KH-8 photography to show the steady technological growth in satellite reconnaissance capability.

On 13 February 1969, Lundahl, his assistant Frank Beck, and DCI Richard Helms were welcomed by Eisenhower in his suite at Walter Reed hospital. Eisenhower, clear-eyed, and ruddy-faced, was as sharp of mind as ever. Beck held briefing boards at the foot of the bed, moving them nearer whenever the President requested a close look (and an opportunity to wield his magnifying glass). Eisenhower shared in the presentation with his old enthusiasm, asking many questions, and remarking his "great satisfaction" over the fine results obtained by the new reconnaissance systems.[127]

Eisenhower must have been equally satisfied with another consideration, not specifically mentioned in the briefing, but implicit in every sentence and photograph. Thanks to the reconnaissance systems which his foresight had nurtured into being, his large vision of "Open Skies"—waved aside so cavalierly by the USSR at Geneva 14 years before—had become a positive reality. Day after day, orbiting satellites were holding denied areas in a steady gaze.

The "haunting concern" of the 1950s had been replaced by the informed response of the 1960s.

Presidential Commendation

Fifteen years later, in August 1984, another Republican President, Ronald Reagan, commented eloquently on GAMBIT's contribution to US intelligence in the following message sent to the National Reconnaissance Office:

Commendation to the GAMBIT Program

When the GAMBIT Program commenced we were in the dawn of the space age. Technologies we now take for granted had to be invented, adapted, and refined to meet the Nation's highest intelligence information needs while exploiting the unknown and hostile medium of space. Through the years you and your team have systematically produced improved satellites providing major increases in both quantity and quality of space photography.

The technology of acquiring high quality pictures from space was perfected by the GAMBIT Program engineers; GAMBIT photographic clarity has yet to be surpassed. Through the years, intelligence gained from these photographs has been essential to myself, my predecessors, and others involved with international policy decisions. These photographs have greatly assisted our arms monitoring initiatives. They have also provided vital knowledge about Soviet and Communist Bloc scientific and technological military developments, which is of paramount importance in determining our defense posture.

A generation of this Nation's youth has grown up unaware that, in large measure, their security was ensured by the dedicated work of your employees. National security interests prohibit me from rewarding you with the public recognition which you so richly deserve. However, rest assured that your accomplishments and contributions are well known and appreciated at the highest levels of our Nation's government.

Appendix A

GAMBIT—Key Contributor to National Security Intelligence

National Intelligence Requirements Management

Prior to examining some of the significant contributions the GAMBIT program made to US national security because of its ability to resolve intelligence questions or problems, it is useful to understand the management structure that provided intelligence requirements guidance over the almost 21-year life span of this remarkably successful program.

In 1963, the Intelligence Community's overhead intelligence requirements were managed by the US Intelligence Board (USIB) through its Committee on Overhead Reconnaissance (COMOR). COMOR was created in August 1960 "for the purpose of providing a focal point for information on, and for the coordinated development of, foreign intelligence requirements for overhead reconnaissance projects and activities of the Government over denied areas."[128] This organization came into being several weeks before the first successful CORONA mission. CIA's James Q. Reber was appointed the first chairman of COMOR.

Prior to 1960, the Ad Hoc Requirements Committee (ARC), which was initially established in 1955 to provide collection guidance for the U-2 program and which subsequently provided requirements guidance for the early CORONA missions, was responsible for national imagery requirements. Reber had been chairman of the ARC from its inception. There also existed, prior to the establishment of COMOR, a USIB Satellite Intelligence Requirements Committee (SIRC) charged with defining required system performance capabilities so that USIB could provide useful guidance to the satellite development agencies. In 1959, the SIRC called for imagery satellite system capabilities of 20-, 5- ▮ -foot ground resolution. Although this stated resolution requirement did not directly influence the development of any US reconnaissance satellite system, it was the first time that a national intelligence entity had attempted to define such a system's capabilities for meeting national intelligence needs.

The responsibilities of the ARC and the SIRC were subsumed into COMOR, when it was established in 1960. The membership of COMOR was comprised of designated officials of the departments and agencies which constituted the Intelligence Community as represented on the USIB: CIA, DIA, NSA, State, Army, Navy, Air Force.

Consultants were appointed from agencies engaged in system development and imagery exploitation—the National Reconnaissance Office (NRO) and the National Photographic Interpretation Center (NPIC).

In July 1967, the Intelligence Community responsibilities for SIGINT and PHOTINT were separated and a new committee, the Committee on Imagery Requirements and Exploitation (COMIREX), under the chairmanship of CIA's Roland S. Inlow, was established to manage expanded responsibilities for overhead imagery collection and exploitation.[129] (See chart below).

Then, in 1975, the Civil Applications Committee (CAC) was established with representatives from the Departments of Commerce, Interior, and Agriculture, the Environmental Protection Agency (EPA), and the Agency for International Development (AID) to exploit satellite imagery for unique civil requirements. COMIREX was charged with overseeing the activities of the CAC to insure national imagery security policies were adhered to in the use of any authorized imagery. Only domestic imagery was eligible for use by CAC agencies, except for AID, where imagery of national disasters such as drought, famine, and flood, was provided to assist the US Government in determining humanitarian aid requirements.

The day-to-day management of the Intelligence Community's collection and exploitation requirements was handled by two COMOR working groups; the Photo Working Group (PWG) was responsible for managing collection requirements and the Exploitation Subcommittee (ExSubCom) was responsible for providing exploitation guidance to the national exploitation centers. With the establishment of COMIREX in 1967, the PWG was changed in name to the Imagery Collection Requirements Subcommittee (ICRS), with its primary functions remaining unchanged.

A major factor that affected the Intelligence Community's interface with GAMBIT Operations was the formal establishment of the National Reconnaissance Organization (NRO) in 1961. All nationally-approved collection requirements were provided to the satellite operator through the NRO's newly-formed Satellite Operations Center (SOC), located in the basement of the Pentagon. Thus, by the time of the first scheduled GAMBIT mission, a completely different satellite operations management concept from that existing during the early CORONA program was in place, albeit its formation was preceded by considerable political infighting between the CIA and the NRO on roles and missions, authorities, and management responsibilities.

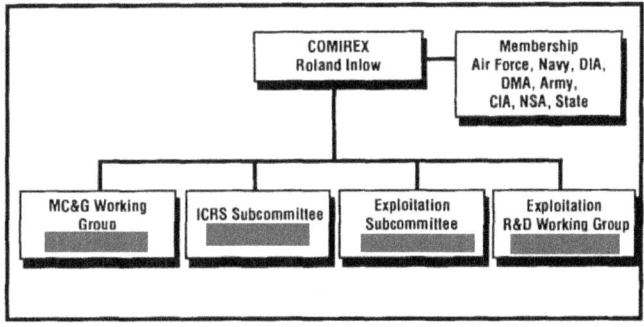

COMIREX Organization Chart

~~SECRET~~
~~NOFORN-ORCON~~

In 1977, all SOC responsibilities for GAMBIT operations were transferred to ▓▓▓▓▓▓▓▓▓▓▓▓. This created an even more cohesive bond between the intelligence requirements manager (COMIREX) and the system operator—NRO ▓▓▓▓▓▓▓▓. In recognition of its broad responsibilities and authorities, ▓▓▓▓▓▓▓▓ was designated, in 1981, an Operating Division (OD-4) under the Secretary of the Air Force ▓▓▓▓▓▓▓▓.

OD-4 played a significant role in the success of the GAMBIT program. Its primary goal was to satisfy intelligence requirements to the maximum extent possible without "breaking" the satellite. To accomplish this it continuously instituted new operational procedures and changes to improve collection capabilities. The basic philosophy was to take any actions possible that would improve requirements-satisfaction without adversely affecting the "health" of the satellite or precluding it from accomplishing its total mission. OD-4, for example, was instrumental in greatly increasing the number of operations GAMBIT could achieve by demonstrating that roll-rates could be increased safely and the time period for each operation significantly reduced. The close working relationship between the COMIREX staff and OD-4, existing throughout the program, also contributed to GAMBIT's success. The following is quoted from a letter to the Director, NRO, from ▓▓▓▓▓▓▓▓ Chairman COMIREX, expressing his appreciation for the outstanding efforts of the OD-4 team in responding to national intelligence tasking.

> I would like to extend my appreciation to you for the outstanding performance of OD-4 KH-8 [GAMBIT-3] system operators in their operation of the final KH-8 Mission recently completed.
>
> This KH-8 team was essential to the successful acquisition and satisfaction of various Intelligence Community collection problems. Additionally, the operators were often requested to provide studies and assistance on unique and/or sensitive requirements within a very short time frame. Their never-flagging spirit, flexibility, and "can do" attitude resulted in a high requirement satisfaction record and the appreciation of the entire Intelligence Community for their efforts.
>
> Again, in keeping with the superior standards established by the National Reconnaissance Office, the KH-8 system team deserves special recognition from all who benefited from their professionalism and expertise.[130]

COMIREX and the NRO (OD-4) constituted two nationally chartered organizations necessary to assure imagery requirements; they were controlled and managed in the best possible national interest: (1) COMIREX to define and prioritize imagery collection, exploitation, and distribution requirements and (2) the NRO to translate collection requirements into command instructions and content and accomplish on-orbit collection, utilizing sophisticated targeting software, weather forecasting, and verification capabilities.

~~SECRET~~

Handle via
BYEMAN-TALENT-KEYHOLE
Control Systems Jointly
BYE 140002-90

GAMBIT Imagery Security Policy

GAMBIT imagery and imagery products were controlled within the TALENT-KEYHOLE (TK) security system. This special system was developed to protect the imagery and imagery-derived products obtained from overhead reconnaissance systems. The TK security system was used primarily by the Intelligence Community for those persons who required certain knowledge about the physical characteristics and performance capabilities of the imaging system but did not require all the system technical and planning data (which was controlled under the BYEMAN security system). Each program also had a TK identifier: for GAMBIT it was KH-7 for the original configuration and KH-8 for GAMBIT-3.

Photointerpreters could be assisted in their analyses by knowing the physical characteristics and performance capabilities of the reconnaissance satellite itself, as well as the operational parameters of each mission. To help the photointerpreters, a special booklet was prepared on the GAMBIT system; this was called the "KH-7 (later KH-8) System Manual" and was security-controlled in the TK system. In addition, operational data unique to each mission were provided to the interpreters, usually covering such matters as vehicle attitude and altitude, solar elevation, and similar matters. Most Intelligence Community members were briefed at the TK level only, rather than at the more comprehensive BYEMAN level; consequently, reference to satellite reconnaissance systems was usually made by using TK designators. Thus GAMBIT was known as KH-7/KH-8 in intelligence circles.

Early in the GAMBIT program, TK clearances were severely restricted, which limited the number of Intelligence Community users who had access to GAMBIT imagery or imagery-derived products. These tight restrictions prevented GAMBIT-derived intelligence from being made available to organizations and activities that had clearly demonstrated a requirement for such intelligence—particularly DoD field elements.

As GAMBIT's collection capabilities steadily improved, it became apparent that the depth and great value of the satellite-derived information made it essential to also make the data available to lower-echelon military and Intelligence Community users outside the TK compartment. Accordingly, in November 1973, President Richard Nixon approved DCI William Colby's recommendation to modify some of the strict security controls on the satellite program imagery. Specifically, the DCI was authorized to remove from TK control (after consultation with the Secretary of Defense) such photographic products as he deemed appropriate, provided that the products so removed were appropriately classified and did not reveal the sensitive technical capabilities of current or future intelligence satellite systems. This authorization resulted in having most of the product (except original format film and almost all of the information derived from it) eligible to meet the requirements of US intelligence users at the Secret level—outside the TK security control system. This action significantly increased the use of intelligence derived from the GAMBIT program. The chairman of COMIREX managed, and continues to manage, the TK security system for the DCI.

~~SECRET~~
Handle via
BYEMAN-TALENT-KEYHOLE
Control Systems Jointly
BYE 140002-90

SECRET
NOFORN-ORCON

The BYEMAN Control System, which manages access to operational and program data on NRP programs, is managed by the NRO; it was unaffected by the DCI's modifications to the TK security system.

Anticipation of Success

As the launching date of the first GAMBIT mission drew near, a sense of excitement and anticipation was apparent throughout the Intelligence Community. Although the NRO had shown simulated GAMBIT imagery to COMOR and other Community organizations, it was difficult for some Community members to actually accept the fact that imagery satellite technology could progress so far and so fast. To improve imagery resolution from as poor as 50-foot ground-resolved-distance (GRD), or worse, on the earliest CORONAs to a projected two feet on GAMBIT, in the space of less than three years, was a spectacular technical achievement and the potential for satisfying priority intelligence requirements was tremendous. The scientific and technical (S&T) organizations of the military services eagerly awaited the high-resolution satellite imagery which would allow them—for the first time—to perform true S&T analysis. Could the imagery really be as good as predicted? The Intelligence Community was not disappointed. With the successful recovery of the first GAMBIT-I/KH-7 mission on 14 July 1963 (GMT), the system's potential—in terms of high-resolution capability—was clearly demonstrated, despite the fact that only three high priority national targets were acquired.

GAMBIT enabled the photointerpreter/analyst to do precise order-of-battle identification and true technical intelligence reporting for the first time using satellite imagery. The capability to enlarge the original negative 100 times—or as much as 2,000 times later in the program—greatly assisted in exploitation of the imagery for technical details. It could be said that the CORONA program removed blinders from the Intelligence Community, with respect to worldwide denied territories, and now GAMBIT provided the required image quality to allow unambiguous intelligence judgments concerning foreign weapons developments, weapons deployment, order of battle, and command-and-control and CC&D information, among other areas of intelligence need where high-resolution imagery was essential. For the first time, the non-photointerpreter would find it easier to believe what he was being told; he could actually identify targets in the imagery.

Requirements Definition Challenge

The development of GAMBIT and its relatively small footprint (5-nm swath, variable length) as compared to CORONA's broad coverage (140-nm swath) presented new challenges for the Community in defining its collection requirements. Whereas CORONA was capable of covering huge areas of denied territory and large numbers of targets in a single operation, GAMBIT would normally be programmed against single targets. Thus GAMBIT was characterized as a "surveillance" system as opposed to CORONA, which was

SECRET
Handle via
BYEMAN-TALENT-KEYHOLE
Control Systems Jointly
BYE 140002-90

a "search" system. (Surveillance is defined as periodic coverage of known installations, including intelligence on equipment and activity associated with both.)

For early GAMBIT missions, requirements were prioritized on a target-by-target basis. Thus, if there were two important targets in close proximity, a decision had to be made as to which had the higher priority, so that the targeting software algorithm could make a proper selection. This led to very vigorous debates among the PWG/COMOR members as to which targets were of highest national importance. Such debate was critical, especially early in the program, when there was usually only a single access to many of the important target areas. A typical early method for resolving priority conflicts was to plot each satellite revolution and its access swath on a large-scale map that also contained all the priority targets. Every revolution was then reviewed by the PWG and, in cases of conflict, the problem was resolved by arbitrary "fine tuning" of target priorities. This was sometimes a long-drawn-out acrimonious process that left all parties dissatisfied. Sometimes it was also an exercise in futility, as the actual orbit achieved after launching often varied significantly from the planned orbit, and a special meeting would be needed to repeat the entire "tuning" process. It was obvious that this procedure could not be used effectively for an extended period, since both the mission duration grew and the number of requirements increased rapidly—from a few hundred targets in 1963 to more than ▮▮▮ by 1984. (See the Surveillance Target Requirements Table for an illustration of numbers of unique targets and requirements for FY-84.)

In solving the requirements-management problem a requirements structure was developed in which sets of like requirements—such as Soviet SS-11 ICBM complexes—were grouped into unique "problem-oriented sets" (POS) and assigned a collection priority based on substantiated intelligence need. Within these POSs, individual requirements that had a high current interest, such as on-going modifications at an ICBM site, could be placed in a special high-current-interest POS with a suitable priority for improving chances of successful collection. NRO simulations conclusively demonstrated to the Community that GAMBIT targeting software, by incorporating such factors as ▮▮▮▮▮▮▮▮▮▮▮▮ and so on, could handle target selection much more effectively than could the manual method of the PWG. This improved methodology for managing target requirements evolved into what became known as the COMIREX Requirements Structure (CRS). The CRS has four principal structural elements: The intelligence problem (IP), collection problem sets (CPS), exploitation problem sets (EPS), and reporting requirements. Together, these structural elements served to integrate imagery collection, exploitation, and reporting requirements, thus making imagery intelligence more responsive to user needs, while, at the same time, facilitating the management of imagery requirements and the evaluation of the imagery intelligence process.

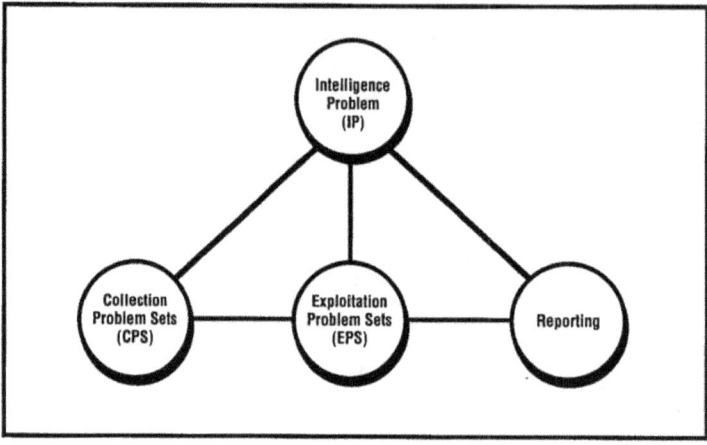

Elements of the COMIREX Requirements Structure

COMIREX Automated Management System (CAMS)

As the definition of intelligence requirements grew more complex and GAMBIT and other NRO imagery satellite programs delivered increasing amounts of imagery, the need for an automated, interactive requirements management system became imperative. Although some computer support to the management of imagery satellite intelligence requirements was available to the community from the earliest days of the CORONA program, all such support was in the form of off-line programs that were useful in mission planning and requirements analysis, but had little utility for near real-time management of requirements *during* the course of a mission. In addition, the Community members could not directly access the national data base to retrieve data on target requirements, imaging attempts, and past coverage. This shortfall was eliminated in 1976, when the COMIREX Automated Management System (CAMS) became operational. For the first time, Intelligence Community members could use a CAMS computer terminal located in their own facility to nominate a collection or exploitation requirement. If the requirement was of a time-sensitive nature, for example, a SIGINT tip-off indicating an on-going ICBM loading exercise at an operational ICBM complex, the COMIREX Staff could take immediate action by directly tasking the NRO to attempt coverage of this requirement on a priority basis. Provided an imagery satellite was on orbit, it could be tasked against such a requirement in a matter of minutes, rather than hours or days.

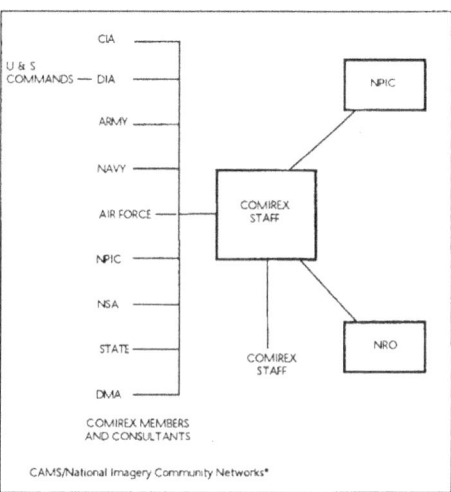

CAMS National Imagery Community Network[131]

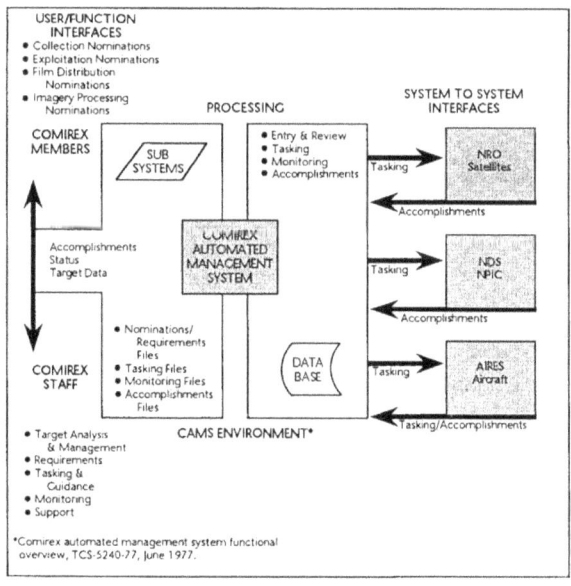

CAMS Environment[132]

National Imagery Exploitation Responsibilities

In 1961, National Security Council Directive (NSCID) No. 8 established responsibility and procedures for the efficient conduct of imagery exploitation in response to national foreign intelligence questions. It created the National Photographic Interpretation Center (NPIC)—for national priority exploitation of satellite imagery—and charged it with providing support services to imagery exploitation organizations in the Washington, D.C., area. The NPIC was also charged with maintaining an up-to-date, consolidated file on imagery-derived target data to serve national and departmental needs. The NSCID directed that imagery exploitation requirements uniquely departmental in nature, for example, DoD studies, were not the direct responsibility of the NPIC; they were to be undertaken by the departments concerned. Those agencies without photointerpretation capabilities, such as the State Department, could call on NPIC to meet its unique readout requirements.

Consistent with NSCID No. 8, a National Tasking Plan for the Exploitation of Multi-Sensor Imagery was promulgated in January 1967. This plan defined the specific roles and responsibilities of Intelligence Community imagery exploitation organizations, which included NPIC, CIA, DIA, Army, Navy, and Air Force, in response to national requirements. National requirements for imagery exploitation by the Intelligence Community were to be developed and managed by COMIREX.

Film-Dissemination Responsibilities

Requirements for overhead imagery dissemination are prescribed by the Exploitation Subcommittee of COMIREX in response to Community needs. Imagery products to be disseminated include film, exploitation data, and printed matter. Additional imagery-related material which must be included with the product are such things as target coverage data, film indexing, camera performance evaluation, mapping, cloud coverage/general weather, requirements satisfaction, and evaluation of overall system performance. The dissemination process is dynamic, continuously supplying data, whether it be on film products, on operational control data and management of a mission underway, on future mission planning data, or on the exploitation of end-products.

National Photographic Interpretation Center (NPIC)

NPIC plays a very important role in the success of overhead imagery programs. The collection of large volumes of high-resolution imagery would serve little purpose without a dedicated and responsive organization to exploit and report on key intelligence information derived from each mission as well as on routine information, such as order-of-battle, on which continuing and long-range intelligence decisions are based. NPIC provided outstanding readout reporting to meet national intelligence exploitation requirements throughout the GAMBIT program. GAMBIT exploitation was divided into three phases:

Phase I. Exploitation and reporting of COMIREX-defined highest-priority targets within 48 hours of receipt of mission film and second-priority targets within five to nine days.

Phase II. Exploitation, reporting, and data base entry of all targets to be accomplished before the launching of the next GAMBIT.

Phase III. Detailed exploitation and reporting of selected targets in support of special (all-source) intelligence reports and studies/estimates at the national intelligence level.

In addressing national exploitation of GAMBIT imagery it is appropriate to mention the first Director of NPIC, Mr. Arthur C. Lundahl. A superb technician in the science of photographic interpretation and photogrammetry, Lundahl used the talents of individuals from diverse disciplines—photointerpretation, photogrammetry, printing and photo-processing, automatic data processing, communication and graphic arts, collateral and analytical research and technical analysis—to extract maximum intelligence from imagery. During his remarkable career he deservedly enjoyed the confidence of Presidents Eisenhower, Kennedy, Johnson, and Nixon, as well as that of senior intelligence managers within the Central Intelligence Agency and the Department of Defense.

Arthur C.
LUNDAHL
Director, NPIC, 1961–1973

Development of the National Imagery Interpretability Rating Scale

As GAMBIT matured, in terms of both quality and quantity of imagery, and as national collection and exploitation requirements rapidly expanded and became more complex, it became apparent that the Community needed a better measure for rating the quality of imagery in terms of satisfying stated requirements. The measure that had been in use since the first successful satellite mission consisted of assessing the pictures as Good, Fair, or Poor. This scale did not give the user or the collector sufficient information on the probability that the imagery would answer a specific intelligence need, such as being able to differentiate between a T-54 and a T-55 tank.

The word "quality" has a different meaning for photoscientists than it does for collection-system engineers. To avoid misunderstanding, a National Imagery Interpretability Rating Scale (known more familiarly by its abbreviation NIIRS, and pronounced "nears") was developed. NIIRS substitutes the phrase "information potential for intelligence purposes" for the word "quality." The purpose of this scale is stated concisely: "to obtain from the photointerpreter a judgment as to the interpretability of an acquired image." As a result of adopting the NIIRS concept in 1972, the Community users acquired a quick, accurate method for assessing whether or not a requirement had been met and, in turn, the collection manager (COMIREX) had a reliable system for continuing tasking of the collector (NRO) and cancelling tasking once the required NIIRS quality had been achieved. The NIIRS rating scale ranged from 0 (which meant that interpretability of the imagery precluded its use for photointerpretation) to 9 (which provided the highest interpretation capability). The following summary includes typical examples for the ten NIIRS categories.

Interpretability Criteria

Rating Category 0

Interpretability of the imagery precludes its use for photointerpretation due to obscuration, degradation, or very poor resolution.

Rating Category 1

Detect the presence of large aircraft at an airfield.

Detect a launching complex at a known missile-test range.

Detect armored/artillery ground forces training areas.

Rating Category 2

Count accurately all large straight-wing aircraft and all-large swept/delta-wing aircraft at an airfield.

Identify a completed Type III-C launching area, within a known ICBM complex, by road pattern/hardstand configuration.

Rating Category 3

Count accurately all straight-wing aircraft; count accurately all swept-wing aircraft; and count accurately all delta-wing aircraft at an airfield.

Detect vehicles/pieces of equipment at a SAM, SSM, or ABM fixed-missile site.

Rating Category 4

Identify a fighter aircraft by type, when singly deployed.

Identify an SA-2 or CSA-I missile by the presence and relative positions of wings and control fins.

Identify trucks at a ground forces installation as cargo, flatbed, or van.

Rating Category 5

Detect the presence of call letters/numbers and alphabetical country designator on the wings of large commercial/cargo aircraft (where alpha-numerics are three feet high or greater).

Identify an SA-1 transporter by overall configuration and details of chassis construction.

Identify a singly deployed tank at a ground forces installation as light or medium/heavy.

Rating Category 6

Identify a FAGOT* or MIDGET* fighter aircraft by canopy configuration when singly deployed. Identify the following missile ground support equipment at a known strategic missile site: warhead/checkout van and fuel/oxidizer transporter.

Rating Category 7

Identify the pitot boom on a FLAGON* fighter aircraft.

Identify a strategic missile transporter/erector (fixed or mobile system) when not in a known missile activity area.

Rating Category 8

Identify on a FISHBED-J* fighter aircraft, the dielectric patch outboard on each wing leading edge and the horizontal tail-plane tip spikes.

* NATO designators for Soviet aircraft.

Identify the VHF antenna on the forward transit support assembly of an SA-4 transporter/launcher.

Rating Category 9

Identify on the appropriate model FISHBED fighter aircraft: wing flap actuator fairings; fairings in after-burner area above horizontal tailplane; pitot boom pitch-and-yaw vanes (when uncovered), and air dump port forward of canopy.

Identify a Mod-3 SA-2 missile by the canards (just aft of nose).

The fully developed GAMBIT-3 was the only overhead imagery system capable of consistently acquiring imagery ▇▇▇▇▇▇▇▇▇▇▇▇▇▇▇▇

Weather Support

Weather support to GAMBIT was provided by a special program-cleared element of the Air Force's Global Weather Central (GWC) facility located at Strategic Air Command Headquarters in Omaha, Nebraska. GWC used inputs from US weather satellites, weather station reports (including those of the Soviet Union), and pilot reports to provide support to US imagery reconnaissance satellite operations.

GWC personnel were also attached to the NRO to provide close interaction in areas of mission scheduling, planning, and on-orbit operations.

Weather support did not play a key role in early GAMBIT operations, since the missions were not film-limited (due to their short on-orbit times). Access to priority targets was limited and the number of target requirements was initially small; however, as mission length extended, weather support played an increasingly important role in contributing to optimum film utilization and, in turn, mission success. Weather support was utilized both in the mission-planning stage (climatological influence on mission scheduling and orbit selection) and on-orbit operations (target weather forecasts to influence target selection and target verification to determine probability that a target had been imaged successfully). As the GAMBIT program matured and became more sophisticated, GWC's weather support developed similarly. A measure of the importance of weather support to GAMBIT was the amount of cloud-free imagery returned using actual weather support versus that which could be expected when using statistical climatology data. For example, in the case of areas of greatest intelligence interest in the Eurasian land mass, climatology showed about 65% of the earth's surface as normally cloud-covered. Therefore, if requirements were programmed disregarding the weather factor, one could expect returns of only about 35 percent cloud-free imagery; however, as missions were extended and weather forecast capabilities were routinely utilized, the cloud-free return averaged 70 percent with a high of approximately 80 percent cloud-free. In effect, weather support to the mature GAMBIT program made it possible to double the amount of cloud-free

WEATHER SATELLITES - PROGRAM 417
OBTAIN COVERAGE OF DENIED LAND AREAS

Program 417 Weather Satellites

SYSTEM ELEMENTS
- BOOSTER —————— THOR / BURNER II
- RCA SPACECRAFT
- ORBIT CONTROL —————— SPIN STABILIZED
- RCA VIDEO SYSTEM
- RCA / BARNES IR SENSOR

PAYLOAD DATA
- VIDEO RESOLUTION —————— 1.0 NM
- CLOUD TOPS MEASURED TO ±1000'
- DATA RECORD AND READOUT

ORBITAL PARAMETERS
- INCLINATION —————— ~98 DEG
- ORBIT —————— ~450 NM CIRCULAR
- LIFETIME —————— ~9 MO

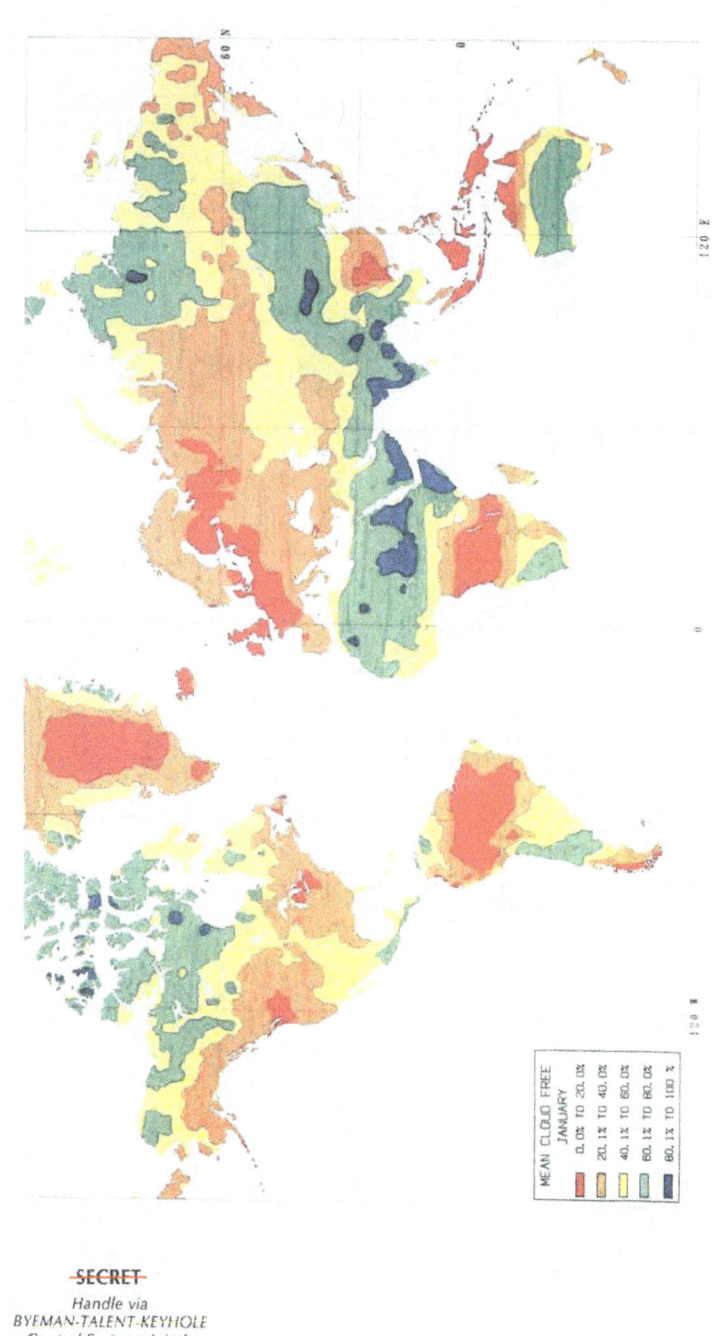

Mean Cloud-Free Areas of the World in January

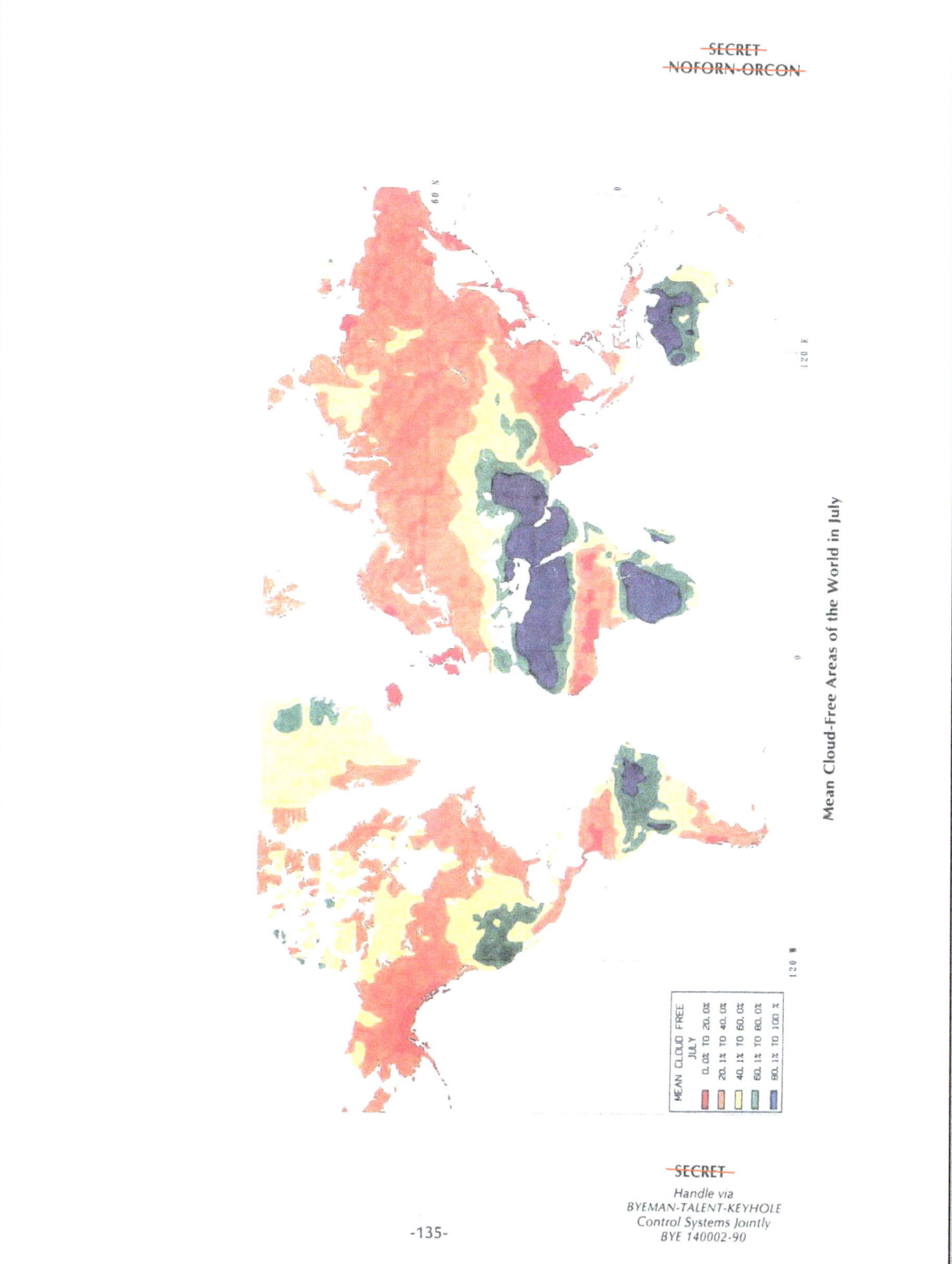

Mean Cloud-Free Areas of the World in July

imagery and, in turn, significantly increased the level of intelligence requirement satisfaction. On the preceding pages are colored charts illustrating the weather problem: the first chart shows the mean cloud-"freeness" for the month of January, the second shows the same data for the month of July. For both months, there is less than a 40-percent chance of observing a target in the primary areas of interest on a given day.

A weather satellite was developed and managed in the "white" world by the Air Force and treated initially as a classified Special Access Required (SAR) Program (Air Force Program 417). In reality, the development was funded by the NRO and was used primarily to support NRO photographic programs, although other military users (including tactical commands) routinely used the data. The program goal was to have a morning satellite, called a "Scout," for forecast purposes with an afternoon Scout for verification. Due to unanticipated mission failures, it was not always possible to have the desired morning-afternoon satellite combination continuously on orbit. Data as current as about three hours old could be applied to GAMBIT weather forecasts when the morning weather satellite was operational.

GAMBIT Intelligence Utility

The CORONA program provided, for the first time in US history, a capability to monitor military and industrial developments over vast areas of the Soviet Union and other denied areas of the world. Although CORONA provided immeasurable contributions to national security, its resolution was not good enough to answer numerous critical intelligence questions, such as those regarding weapons development, that the United States needed to guide counter weapons development. Nor could it provide the image quality the scientific and technical (S&T) intelligence organizations required to do true S&T analysis. GAMBIT aptly filled this high-resolution need and, by the end of the program, was routinely collecting imagery of ▓▓▓▓ ground resolved distance (GRD), or better. The following Table illustrates the wide range of target categories and geographic areas that GAMBIT was able to photograph routinely. Although the data shown are for a single GAMBIT mission, the numbers and distribution are typical for the mature GAMBIT system.

It has been asserted by individuals responsible for major weapon system developments that one of the greatest contributions of GAMBIT was the billions of dollars it helped to save in US weapons development. For the first time, US personnel had enough detailed information and accurate mensuration data to develop engineering drawings on foreign weapons capabilities. This facilitated the design of cost-effective counter weapons systems; it was no longer necessary to design against a "worst case" possibility.

In 1981, the NPIC identified a number of key historical events for which GAMBIT provided significant intelligence information.[133] They were:

a. The Soviet strategic submarine story with emphasis on the Y- and D-class submarines (1969–75).
b. The Soviet Union vs. United States race to the moon, with emphasis on Launching Complex J at Tyuratam Missile Test Center (1965–72).
c. The Soviet ground order-of-battle story to include:
    (1) Unit reporting (1973–present).
    (2) The Sino-Soviet border (1965–present).
d. The Soviet strategic missile story, with emphasis on SS-9 developments (1964–71).
e. Variations in aircraft (a number of examples exist although the Bear-F variants were cited) (1969–present).
f. Communication vehicles/equipment first identified by GAMBIT. This may ultimately include big radars like DOG HOUSE, HEN HOUSE, and over-horizon detectors (1963–present).
g. The Anti-Ballistic-Missile vs. Long-Range SAM (SA-5) Controversy (1963–69).
h. The development of large solid-propellant motors for strategic missiles at Pavlograd (1969–present).
i. The SS-16/20 mobile-missile story from birth to deployment (1972–present).
j. The Caspian Sea Monster story (1967–present).
k. The evaluation of Soviet reconnaissance programs, based on the sighting of Seiman Stars (1974–77).
l. The development of Soviet camouflage, concealment, and deception techniques (1963–present).
m. The construction of suspect advanced weapons-related facilities in China (1967–72).
n. The identification of the Chinese planar-array antenna (1974).
o.
p.

Satisfaction of Major Intelligence Requirements

The following photographs are examples of GAMBIT's continuous contribution to satisfying major intelligence requirements during the 1963-84 timeframe.

● *Soviet Ballistic-Missile Submarine (SSBN) Production.* In the 1960s and 1970s the Soviet push for nuclear supremacy was of great concern to US leaders. One of the greatest concerns was the construction rate and operational capabilities of Soviet missile-launching submarines known as SSBNs. GAMBIT imagery could closely monitor the production rate of various SSBN models, as well as provide technical intelligence details on numbers and types of propellers, number and size of missile tubes, hull construction (particularly important in designing the type of weapon required to sink it), surfaced and submerged displacement, and so on. Operational training and deployment tactics could be monitored, since GAMBIT imagery could identify specific submarines by unique hull markings. The following graphics are typical examples of GAMBIT imagery associated with the SSBN problem. Graphic C clearly demonstrates the high-resolution qualities of GAMBIT, when compared to the HEXAGON/KH-9 search system. Both images have been enlarged 30 times. Graphic E shows GAMBIT's capability to detect deception attempts easily. The imagery of 27 June 1974 identified ▮▮▮▮▮▮▮▮▮▮▮▮▮▮ imagery of 3 and 4 July 1974 is from the second bucket of the mission.

(A) Delta-Class Hull-Staging Area at Severodvinsk—20 May 1973

● *Missile Test Ranges.* GAMBIT provided insight into Soviet and Chinese missile and space development and operational procedures by observing major missile test facilities on a regular basis. Indications of new ICBM, IRBM, ABM, SAM, mobile ballistic-missile systems, or space-launching vehicles were routinely detected at these ranges. This information was vitally important to strategic planners, as well as to representatives at SALT discussions and other arms negotiations. Graphics A through E are examples of test-range imagery. Graphic C clearly shows the extensive damage resulting from catastrophic failure of Soviet launching attempt on the Tyuratam J-Pad on 3 July 1969. This was just 17 days before the US launching of Apollo-11 which involved the first manned excursion to the lunar surface. As a result of this accident and the US success, the Soviets abandoned further attempts of manned exploration of the moon.

(A) Tyuratam Missile Test Center, Complex A—16 Mar 1968

(D) Shuanchengtzu Missile Test Center, PRC, Complex A—29 May 1967

● *Operational ICBM Complexes.* The Strategic Arms Limitations Treaties were made feasible by US capabilities to monitor Soviet strategic weapon deployments. GAMBIT photography was a key in the S&T analysis of Soviet and Chinese ICBM complexes. Graphics A and B illustrate GAMBIT's capability to monitor construction of new facilities and provide technical information on such important elements as command-and-control bunkers and silo hardness. Graphic C provides an illustration of GAMBIT's capability to monitor site readiness prior to silo hardening of all ICBM complexes. The Soviets attempted camouflage and deception at many operational bases; these were easily detected on GAMBIT imagery as illustrated in Graphics D and E.

(A) Itatka ICBM SS-7 Soft Site—18 Apr 1968

● *Soviet Ground-Force Divisions.* GAMBIT imagery was used extensively in the mid-1960s to help resolve a dispute within the Intelligence Community concerning the size of the Soviet ground-force divisions. The dispute grew out of remarks by Premier Nikita Khrushchev that, as a result of Soviet ICBM deployment, Soviet armed forces could be reduced from 3.6 million to 2.4 million. At the same time the Soviet Union claimed 80 combat-ready divisions. These two claims were incompatible if all Soviet divisions were manned and equipped at the same level as those that could be observed by high-resolution aircraft photography in East Germany. Defense Secretary Robert McNamara ordered an imagery study, known as Operation MILOB, to resolve this paradox. In the resulting concentrated examination of available imagery of the entire Soviet Belorussian Military District, including over 5,000 prints of GAMBIT imagery, photointerpreters were able to determine accurately the amount of storage area for ground-force equipment. A major finding was that far fewer pieces of military equipment existed in Belorussia than had previously been estimated; this confirmed that ground-force divisions within Soviet borders were smaller than those outside its borders and, therefore, that the Soviet Union had fewer ground troops and equipment than previous National Intelligence Estimates (NIEs) had assumed. The following graphic illustrates GAMBIT's capability to do order-of-battle counts at Soviet ground-force installations.

Borisov Army Barracks—15 Aug 1968

Scientific and Technical Intelligence

GAMBIT's contributions to scientific and technical (S&T) Intelligence were unsurpassed. The mature system produced examples ▓▓▓▓▓▓▓▓▓▓▓▓▓▓▓▓▓▓▓▓▓▓▓▓▓▓▓▓▓▓▓▓. Furthermore, it exhibited excellent mensuration capabilities, allowing the S&T photointerpreter to perform accurate measurements on foreign weapons systems, command and control and control systems and research and development hardware. As noted earlier, this saved the US Government significant defense funds in weapon development, as well as allowing accurate intelligence judgments of Soviet (and other countries') offensive and defensive capabilities.

The following ten graphics are illustrative of GAMBIT's high-quality imaging capability.

● *Soviet Phased-Array Radars.* Construction of the Soviet Phased-Array Radars relative to the ABM and ASAT questions was of high interest to the United States policymakers. High resolution imagery during the construction phase was especially important for analyses of system capabilities.

Abalakovo Phased-Array Radar

● *Determination of Silo Hardness.* Imagery of new, modern solid-propellant silos at the Plesetsk Missile Test Range, obtained by the GAMBIT system during the construction phase of new Soviet ICBM weapon systems, produced important information on silo hardness, launching design, as well as intended weapon systems. Such data were invaluable to US SALT/START negotiators, as well as for strategic targeting planners.

Plesetsk ICBM Silos

● *Soviet Aircraft Carrier at Nikolayev.* US policymakers and defense planners were able to monitor and measure construction, from the laying of the keel through the fitting-out process, of this Soviet aircraft carrier under construction at Nikolayev Shipyard.

Soviet Aircraft Carrier Construction at Nikolayev Shipyard

- *Soviet Delta-Class Submarine.* GAMBIT imagery of a Delta-class submarine at Severodvinsk shipyard with its missile tubes open made it possible to measure the numbers and types of missile tubes. This provided accurate assessments of the submarine's weapon system, indicative of its strategic threat.

Delta-Class Submarine With Missile Tubes Open at Severodvinsk

Soviet Deep-Space Radar. This 80X enlargement of the Soviet Deep-Space Radar Tracking facility at Yevpatoriyo, USSR, illustrates GAMBIT's capability to image electronic equipment in great detail.

Yevpatoriyo Deep-Space Radar-Tracking Facility

● *Soviet AWACS Aircraft.* GAMBIT imagery of this new AWACS aircraft made it possible to measure its radar and other antennas thereby providing information for judging its mission and capabilities.

New Soviet AWACS Aircraft

● *Typhoon SLBN Submarine.* This image of the Typhoon submarine at Severodvinsk illustrates the level of detail achievable by high-quality GAMBIT imagery.

Typhoon Submarine at Severodvinsk

● *The Dual-Platen Camera.* This illustration shows simultaneous *GAMBIT* imagery of a Soviet communications satellite station using two different film emulsions to achieve specific S&T objectives.

Type SO-112 Film

Type SO-209 Film

● *Soviet BLACKJACK Bomber.* Early KH-8 photography of an advanced bomber at the Kazan Airframe Plant was originally designated the Kaz-A. Later, it was given the NATO designator BLACKJACK.

Kaz-A BLACKJACK Bomber at Kazan Airframe Plant

● *Golden Gate Bridge.* A 40X enlargement of one of the towers of the Golden Gate Bridge in San Francisco is included to allow the reader to relate GAMBIT photo-quality to a familiar object.

A Tower of Golden Gate Bridge as Imaged by GAMBIT

● *Soviet Enigmas.* Throughout the GAMBIT Program, its high-resolution capabilities were called upon frequently to resolve perplexing intelligence questions—sometimes with success, sometimes not. An example is the Soviet development of very large surface-effect vehicles in the Caspian Sea. GAMBIT's high-resolution capability provided accurate mensuration of many different versions and allowed US photointerpreters to estimate potential capabilities. The question of whether or not the "Caspian Sea Monster" has a military role has not been determined to this date. See illustrations A and B. Graphic B demonstrates the ability to do engineering drawings from GAMBIT imagery. It is interesting to note that the last clear image (graphic C) of the "Caspian Sea Monster" was made on 11 August 1984 by the final flight in the GAMBIT series.

(A) 'Caspian Sea Monster'—19 Mar 1968

(B) Drawing of 'Monster' from GAMBIT Imagery

(C) Last GAMBIT Photo of 'Monster'—11 Aug 1984

(B) Infrared Netting at SS-20 Mobile Base at Gresk—12 Apr 1981

● *Weapons Model Construction.* High-quality GAMBIT/KH-7 and KH-8 imagery has been used extensively to construct three-dimensional models of foreign weapons systems and facilities. These models have been used to brief senior policy-level personnel and assist engineers in determining weapons system characteristics, such as hardness of Soviet ICBM silos. The following graphic is an example of one such model.

Model of a Soviet Type-IIIC ICBM Site

● *Color Imagery.* Various color, false-color, and infrared (IR) films were used throughout most of the GAMBIT program. Although some unique intelligence was acquired and some of the imagery is spectacularly impressive, the consensus of the Community was that color never proved to be a major source of additional intelligence. The thicker film emulsion and coarser grain characteristics of color sometimes degraded intelligence utility because of poorer resolution/NIIRS values. This disadvantage was overcome somewhat later in the GAMBIT program, with the introduction of the dual-platen camera, starting on improved GAMBIT mission No. 4348, in March 1977, when high-resolution black-and-white film could be spooled on the nine-inch film supply and special-purpose films spooled on the 5-inch film supply. Thus color and black-and-white imagery were obtained simultaneously. This arrangement was only partially successful, because it created a very difficult film-management process to assure that desired special film was available on the 5-inch platen when needed. The following examples of GAMBIT color and IR imagery show the Berenzniki Chemical Combine in the USSR and a rail-to-road transfer point near Yurya.

Berenzniki Chemical Combine

Yurya Rail-to-Road Transfer Point With Infrared-Reflectant Netting

SECRET
NOFORN-ORCON

● *Mapping, Charting, and Geodesy Contributions.* Some mention of GAMBIT's ability to satisfy MC&G requirements is worth noting. Although accorded little publicity throughout the program, GAMBIT did satisfy Defense Mapping Agency (DMA) requirements for high-resolution imagery of foreign urban areas, airfields, ports, and harbors; accurate updated DoD maps could be compiled. Literally thousands of such specific requirements were satisfied by GAMBIT. The "satisfaction level" was high because the requirement existed worldwide and was usually not in conflict with priority intelligence requirements. Also, in dense Sino-Soviet target areas, MC&G requirements could frequently "piggy-back" on higher priority intelligence requirements. The final GAMBIT mission provided an illustration of the extensiveness of DMA's MC&G requirements. Of the ▓▓▓▓ requirements tasked to this GAMBIT mission, ▓▓▓▓ percent were MC&G requirements.

~~SECRET~~
~~NOFORN-ORCON~~

Appendix B

A CORONA Summary*

The CORONA Program was approved for development by President Eisenhower on 7 February 1958. At White House direction, the program was organized under the joint leadership of CIA's Richard M. Bissell, Jr., and US Air Force Brig. Gen. Osmund J. Ritland. CORONA was a breakout from a larger satellite reconnaissance development called WS-117L, which was being conducted at the Air Force Ballistic Missile Division (AFBMD) in Inglewood, California. A portion of WS-117L, called Discoverer, was the precursor of, and cover for, CORONA.

The AFBMD was responsible for all hardware required for CORONA—except the camera—and, additionally, for providing launching, tracking, and recovery facilities to the program. The CIA funded the camera development and reentry vehicle procurement, provided security supervision for the "black" aspects of the program, and defined its covert objectives.

The Lockheed Missile and Space Division (under contract to both the CIA and BMD) was to integrate all equipment, develop the upper (spacecraft) stage, and furnish leadership in testing, launching, and on-orbit control operations. Itek developed the camera, General Electric built the recovery capsule, and Douglas furnished the Thor boosters.

CORONA security kept the program "black." This was not hard to do, since, to the uncleared world, CORONA could be presented as the old Discoverer—a technological program for exploring the space environment and for pioneering assistance to later satellites. The CORONA launching site would be Vandenberg AFB; its control station would be at Sunnyvale, and recovery ships and aircraft would work out of Oahu.

CORONA No. 1 was launched on 28 February 1957, purely as a test-flight. In a subsequent series of eleven flights, extending to August 1960, there were no successes. Flight No. 13, a diagnostic flight, carrying only test instrumentation, was recovered by water-pickup on 12 August 1960. But the first actual success—with "success" measured in terms of exposed film delivered—was flight No. 14, air-recovered on 18 August 1960.

In the first two years of operation, dating from 18 August 1960, 48 photographic missions were attempted with 19 "true" successes. The original camera, retrospectively called KH-1, produced nominal resolutions of 40 feet; with improvement in cameras, models known as KH-2 and KH-3, as well as film, resolutions began to move below 10 feet. There was continual improvement in the CORONA system. A stereoscopic arrangement, called CORONA-M and also known as KH-4, was introduced in 1962. In 1963, the CORONA-J, also known as KH-4A, entered the inventory. It was capable of carrying 15,000 feet of film in each of *two* re-entry capsules. The final improvement was the constant-rotator camera, the KH-4B, which achieved resolutions as small as six feet at nadir.

*See also F.C.E. Oder, James C. Fitzpatrick, Paul E. Worthman, The CORONA Story, December 1988, BYE 140001-88.

~~SECRET~~

Handle via
BYEMAN-TALENT-KEYHOLE
Control Systems Jointly
BYE 140002-90

CORONA's life span, as a program, was 12 years and covered 145 launchings. Ground resolutions of 6-10 feet were eventually achieved. By 1970, CORONA could remain in orbit for 19 days, make operational responses to cloud-cover, provide accurate mapping information, and return coverages as large as 8,400,000 nm². The final cost of an average mission was ███████.

The Intelligence Community described CORONA's contribution to its resources as "virtually immeasurable."

Appendix C

Leningrad and LANYARD: Search for the GRIFFON[134]

During the late summer of 1961, NPIC photointerpreters, examining imagery obtained by a CORONA satellite (mission No. 9017, launched in June 1961), discovered clearing and site preparation work near Leningrad. This construction resembled prototype structures photographed early in 1960 during a U-2 overflight of Saryshagan Missile Test Center, where the Soviet anti-ballistic missile (ABM) effort was headquartered. Work on this Leningrad system continued throughout 1962 and eventually involved three sites with nearly 30 launchers. A similar installation was seen in CORONA imagery of Tallinn, Estonia. The Intelligence Community debated the mission of this Leningrad system (assigned the NATO designator GRIFFON) but had precious little high-resolution imagery on which to base its estimates. The Air Force believed it to be an ABM system, CIA and the Army thought it was designed to interdict high-flying US bombers such as the B-52 and B-58.

In 1961, CORONA's KH-2 imagery could resolve no objects smaller than 10-15 feet on a side at nadir; consequently, photo-interpreters could not distinguish between the GRIFFON SA-5 missile and the air-to-air GUIDELINE SA-2, which were approximately the same length. Meanwhile, a new H-configuration was seen for the radars at GRIFFON sites around Leningrad. Again, CORONA imagery was such that interpreters could not determine the type of radar antennas being installed at these H-sites, information that would be important in determining the ELINT parameters of the radar.

By early 1962, Secretary of Defense Robert S. McNamara, already confronted with a worsening situation in Southeast Asia, now had to countenance the possibility of undertaking the expensive development of an ABM system. Before taking such a step, McNamara urged DCI John A. McCone to get better pictures so that NPIC's photointerpreters could be more positive in their identification of the Leningrad weapon system. In his turn, McCone urged DNRO Joseph Charyk to do everything possible to obtain high-resolution photographs of the Leningrad system, including speeding up the launching of the GAMBIT satellite (with its 77-inch focal-length camera). Charyk, however, realized that it would not be possible to launch the GAMBIT system before mid-1963. Consequently, in April 1962, he signed an agreement with CIA's Deputy Director for Research, Herbert "Pete" Scoville, Jr., for a joint Air Force-CIA "crash" effort to provide an interim spotting satellite using part of the proven CORONA system and a high-resolution E-5 camera which had been developed by Itek Corporation for the moribund Samos program.

This hybrid effort was known as Project LANYARD and its camera was designated the KH-6. LANYARD was to be overseen by CIA's West Coast Contract Office and, like the ARGON mapping-camera (KH-5) effort for the Army Map Service, was to come under the CORONA security cloak. Consequently, the contractors working for the "black" Air Force were not witting of the project.

It was hoped that the LANYARD effort could adapt existing and proven launching and recovery systems to accommodate the E-5/KH-6 camera. This device, with its 66-inch focal length and f/6.0 optics, was expected to provide a resolution of 5 to 6 feet while photographing a swath about 40 miles wide. The LANYARD camera not only had a focal length 42 inches longer than CORONA, it also used bigger film, 127-mm (5-inch) compared with CORONA's 70-mm (2.75-inch) film.

Bigger Spacecraft, New Booster, Roll-Joint Needed

Although the Itek E-5 camera had already been built and the CORONA/Agena spacecraft and film-return system were fully operational, there remained a considerable problem in mating the camera to the existing system. In August 1962, LMSC undertook to enlarge CORONA's 9-foot-long spacecraft so it could accommodate the E-5/KH-6 camera. The new enclosure was 14 feet long.

The heavier LANYARD payload also required more thrust to put it into a polar orbit and Douglas Aircraft Company began work in late 1962 to develop a more powerful Thor rocket for launching this interim "spotting" satellite. The new booster was known as the thrust-augmented Thor, or TAT, and consisted of a standard Thor missile to which were strapped three solid-propellant rockets (manufactured by Thiokol Corporation) which could be jettisoned after firing. The new TAT configuration was first tested on 28 February 1963, when it was used to launch CORONA mission No. 9052. Unfortunately, one of the strap-on boosters failed to separate and the entire mission was destroyed 100 seconds after launching.

In developing LANYARD, LMSC designed and built a "roll-joint" which permitted the camera segment of the spacecraft to rotate up to 30 degrees from the vertical while attached to the Agena-B. The roll-joint, a planetary gear arrangement, made it possible to point the E-5/KH-6 camera at off-axis targets to either side of nadir. During the roll operations, the Agena-B maintained X-Y-Z-axis stability for the entire orbiting platform. The limitation of the LANYARD roll-joint was that it would provide only 100 stereo pairs of pictures of selected targets during a single mission. (This was only 25-30 percent of the number of stereo-pairs that the GAMBIT system hoped to produce with its orbital-control vehicle.)

The first LANYARD satellite, mission No. 8001, was launched on 18 March 1963. The TAT worked smoothly but the satellite failed to go into orbit, because of a second-stage Agena-B malfunction. A second flight, mission No. 8002, on 18 May 1963, went into orbit and its payload was successfully returned to earth, but the E-5/KH-6 camera had failed and no pictures had been taken.

The P-Camera Experiment

Meanwhile, the pressure from within the Intelligence Community for high-resolution imagery of the Leningrad system had reached a point where the Directors of Programs A and B were literally clutching at straws. On 22 April 1963, DCI McCone flew to Boston and persuaded Dr. Edward M. Purcell of Harvard, an original member of Edwin Land's 1954 TCP Intelligence Panel, to chair a panel to survey the future of reconnaissance satellites and consider methods for improving their imagery.

During this meeting, McCone mentioned to Purcell the problem of obtaining imagery of the Leningrad site. The Nobel-prize-winning physicist suggested a quick-and-dirty method of obtaining such imagery: put a telescope and strip camera in a CORONA satellite and photograph the Leningrad target. He thought this might be done with a minimum expenditure.

The suggestion was passed along to the CORONA Program Office, which approached Itek Corporation with the idea. Using off-the-shelf parts, Itek built a 240-inch Cassegrain telescope, using "folded optics," and coupled it with a 127-mm strip camera. This came to be known as the P-(for Purcell)-camera experiment. Meanwhile, in California, modifications were made to a standard CORONA-M spacecraft. By using vacant space within the film-transport area of the spacecraft, Lockheed engineers were able to install a dummy unit, the same size and weight as the P-camera. They also cut an optical port which, like the optical ports for the KH-4 camera, was provided with a protective door. After launching and orbital insertion, these doors were blown off with small pyrotechnic devices.

On 12 June 1963, CORONA mission No. 9054 was sent aloft with its normal MURAL/KH-4 camera payload plus the dummy P-camera. The CORONA program managers and engineers hoped to determine: (1) if the P-camera would fit into the payload area without disrupting the functions of the MURAL camera; (2) if the TAT could boost this heavier load into orbit; and (3) if the Agena-B's on-orbit control systems could stabilize the spacecraft with this second device inside. The SRV was deorbited on 16 June after a normal mission. The KH-4 camera had exposed its full load of film and there was no apparent difficulty in maintaining spacecraft stability.

Then, on 26 June 1963, CORONA mission No. 9056 was orbited, with the one-and-only P-camera on board, along with a standard MURAL camera. Everyone was anxious to see the results of this experiment and hoped that more could be learned about the Leningrad system. On the CORONA satellite's first engineering pass over the Satellite Control Facility at Sunnyvale, the spacecraft's housekeeping telemetry indicated that the door covering the P-camera's optical port had not blown off. Lt. Col. Vernard Webb, CIA's chief of satellite operations on the West Coast, was hopeful that this was faulty telemetry. He ordered the camera turned on during the next pass over Leningrad. The SRV was deorbited on 30 June and a normal recovery was made. When the film was developed the P-camera's film was blank, proving that the optical-port door had not blown off.

SECRET
Handle via
BYEMAN-TALENT-KEYHOLE
Control Systems Jointly
BYE 140002-90

A little more than two weeks after the first successful GAMBIT launching, a third LANYARD system, mission No. 8003, was sent into polar orbit from Vandenberg AFB on 30 July 1963. The E-5/KH-6 camera failed during the 23d orbit, after exposing only 25 percent of the film. The payload was recovered successfully on 1 August. Much of the LANYARD imagery was degraded by focus aberrations; nevertheless, some useful photography in the 5.5-foot range was obtained, but there was no imagery of the Leningrad SA-5 sites. In all, five LANYARD systems were assembled, three were launched, but only one was partially successful. DNRO Brockway McMillan cancelled Project LANYARD shortly after the second GAMBIT-1 satellite brought back usable photography on 8 September 1963.

LANYARD's roll-joint, however, proved to be the saving technology for the GAMBIT program when General Electric's concept for an orbital-control vehicle encountered difficulties early in the program. The LANYARD roll-joint was transferred into the GAMBIT effort in early 1963 and remained a vital part of the program for more than 20 years.

As for the SA-5 installations around Tallinn, they remained unphotographed by high-resolution satellites until 1965, but not for lack of trying. The area of the Soviet Union around the Gulf of Finland, which includes Leningrad and Tallinn, is notorious for its cloudy weather. Although there are bright days, when the sun is filtered through high cloud or low-lying mist, the Gulf of Finland is cloud-covered, as far as satellite cameras are concerned, 95–97 percent of the time. The original 1961 imagery of the SA-5 sites was more fluke than skill, and, despite the efforts of NRO planners, the Tallinn sites remained obscure until seen by GAMBIT mission No. 16 and its KH-7 camera from 13 to 16 March 1965.

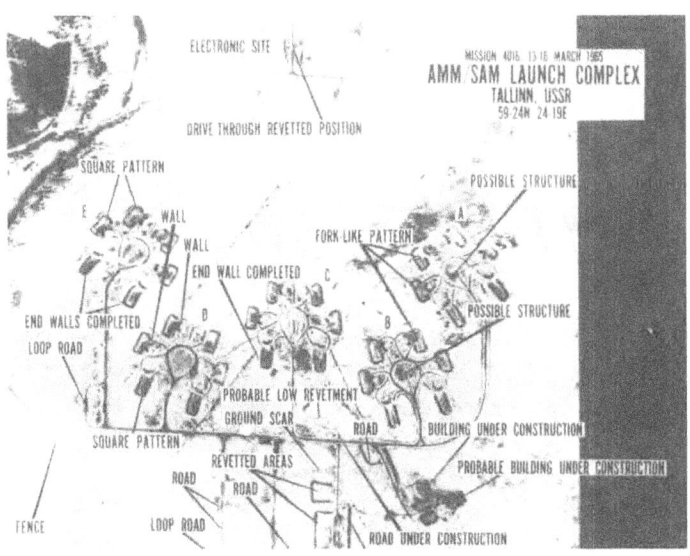

References

1. John S.D. Eisenhower, *Strictly Personal,* Doubleday and Co., Inc., Garden City, NY, 1974, p. 609.
2. James R. Killian, Jr., *Sputnik, Scientists and Eisenhower: A Memoir of the First Special Assistant to the President for Science and Technology,* The MIT Press, Cambridge, MA, 1977, p. 68.
3. George B. Kistiakowsky, *A Scientist at the White House,* Harvard University Press, Cambridge, MA, 1976, p. xxi.
4. Dwight D. Eisenhower, *Mandate for Change, 1953-1956,* Doubleday and Co., Inc., Garden City, NY, 1963, pp. 77-81.
5. *Ibid.,* p. 1.
6. W.W. Rostow, *Open Skies, Eisenhower's Proposal of July 21,* 1955, University of Texas University Press, Austin, TX, 1982, p. xiii.
7. Eisenhower, *op.cit.,* p. 522.
8. *Ibid.,* p. 483.
9. For more information on CORONA, see Appendix B.
10. Named for a Grecian island. Almost instantly (and erroneously) elevated to acronymic status by the press, as representing "Satellite and Missile Observation System."
11. Kistiakowsky, *op.cit.,* p. 141.
12. *Ibid.,* p. 192.
13. *Ibid.,* p. 246.
14. *Ibid.,* p. 397.
15. Letter, A.B. Simmons to J.V. Charyk, 22 Jul 60, no subj.
16. Memo, Exec. Sec. USIB for USIB, 5 Jul 60, USIB-D-33.6/8, USIB/S files, Secret.
17. Minutes of USIB Meeting, 9 Aug 60, USIB-M-111 Approving DCID No. 2/7, Based on USIB-D41.2/1 (revised) 28 Jul 60. USIB files, Secret.
18. Memo for Record, BGen R.D. Curtin, 10 Aug 60.
19. Letter, F.G. Foster, EKC, to BMD attn. Maj ███████ 13 Aug 60, no subj.
20. Memo for Record, BGen R.E. Greer, Dir S/P, 20 Sep 60, subj: "SAMOS Program Meeting with SAFUS."
21. Memo for Record, Col. P.J. Heran, 8 Nov 60, subj: "Trip Report to HQ USAF" (2-4 Nov 60); TWX SAFMS-99153, SAFUS to BMD, 4 Nov 60.
22. TWX AFMPP-WS-2-96055, USAF to AMC 24 Oct 60; TWX DPL3416 SAC to ARDC/BMD 17 Nov 60; ARDC Ops Order 60-1, 23 Nov 60.
23. Memo (Secret) J.V. Charyk, SAFUS, to C/S USAF, 6 Dec 60; subj: "Basic Policy Concerning Samos."
24. Memo (Secret-Special Handling), J.V. Charyk, SAFUS, to C/S, USAF, 6 Dec 60, subj: "Basic Policy Concerning Samos."
25. Kistiakowsky, *op. cit.,* p. 300.
26. Memo for Record, Col. P.J. Heran, 8 Nov 60, op. cit.
27. Memo (not sent), BGen R.E. Greer, Dir/Samos Proj., to SAFUS, approx 28 Dec 60, Subj: Progress Report Samos.
28. *Ibid.*
29. "Satellite Reconnaissance Plan," 3 Apr 61, SAFMS, in SAFSP-3 files.
30. *Ibid.*
31. *Ibid.*
32. Robert Perry, *A History of Satellite Reconnaissance, Vol IIIA—GAMBIT,* Jan 74, pp. 23-24.
33. "Subsystem Engineering Analysis Report," Eastman-Kodak Co. 7th Issue, 12 Aug 65.
34. Memo, "Air Retrieval for Program 483A," W.F. Sampson, Aerospace Corp to Col. Q.A. Riepe, Dir. 206 Prog, 19 Jan 62.
35. Msg, ███████ 2280, approx 3 Aug 62.
36. TWX, SAFSS-DIR-M-2082 to Dir/SP, 24 Aug 62.
37. "206 Program Report," GE, 1967.
38. Memo, BGen J.L. Martin, Jr., Dir/SP to DNRO, 29 Aug 67, Subj: "Summary Analysis of Program 206 (GAMBIT)."

References (continued)

39. TWX SAFSS-DIR-M-2095, J.V. Charyk, SAFUS, to M/Gen R.E. Greer, Dir/SP, 30 Oct 62.
40. As noted earlier, the E-6 program was cancelled on 31 Jan 63.
41. Greer interview by R. Perry, 22 Nov 62.
42. For background information on LANYARD, see Appendix C.
43. Memo for Record, Col. W.G. King, Subj: "Development of Roll-Joint Capability," 28 Jan 63.
44. TWX, SAFSS-1-M-2129 to Greer, 30 Nov 62.
45. TWX SAFSS-1-M-2138, SAFSS to SAFSP, 19 Dec 62.
46. The first GAMBIT hardware on the VAFB launching pad was an engineering model, acting as a "stand-in" for rehearsals of (Atlas) booster check-out procedures. On 11 May 1963, problems with a faulty booster valve and with the fuel-loading sequence caused the Atlas to lose its critically-essential internal pressure and collapse. The episode occurred during what was solely a static check-out; no flight phase was intended or possible, in this mode.
47. Dr. Flax was the only DNRO who was, in a "white" capacity, Assistant Secretary of the Air Force (R&D).
48. Interview, MGen J.L. Martin, Jr., USAF (Ret.), 26 Mar 86.
49. Ibid.
50. Ibid.
51. Interview, Col. F.S. Buzard, (USAF-Ret), Jul 86.
52. Martin interview, op. cit.
53. ███████████████████████████████ WADC Historical Studies, 1955.
54. Chuck Yeager and Lee Janos, Yeager, An Autobiography, Bantam Books, 1985, pp. 179-80.
55. MGen John L. Martin, Jr., "A Specialized Incentive Contract Structure for Satellite Projects," Ref. No. SP142866, 18 Apr 69.
56. MGen J.L. Martin, Jr., written communication, 13 Mar 87.
57. Interview, Capt. Frank B. Gorman (USN-Ret) Nov 86.
58. Memo MGen R.E. Greer, Dir/SP to Col. W.G. King, Dir/GAMBIT Ofc, 13 Dec 62; Subj: G^3.
59. "Preliminary Development Plan for G^3," Jan 1984, SAFSP/████.
60. Memo, MGen R.E. Greer, Dir/SP, to Col. W.G. King, Dir GAMBIT Ofc, 2 Jan 64, Subj: G^3.
61. "Preliminary Development Plan for G^3," op. cit.
62. Ibid.
63. Ibid.
64. Ibid.
65. Quarterly Progress Report (QPR) to DNRO, 30 Sep 64.
66. Interview, Leslie Mitchell, Eastman Kodak Co., Oct 86.
67. QPR, 30 Jun 65.
68. "Program 110 Status Book" SAFSP ████. 1972 BYE 94162-72.
69. PET Report, Mission 4301/6, updated.
70. "To be read out, a photograph must represent the confluence of several aspects of success. The camera must have operated correctly, causing no perturbations which would make the negative unreadable. Such perturbations could include uncompensated smear, incorrect focus, faulty compensation for thermal effects, solar or terrestrial flare, and (occasionally) degraded optics. The film had to be unmarred either by faulty manufacturing or by scratches caused by film transport or take up. All of these factors were nominally controllable in manufacture and checkout. The major cause for unreadable film, however, was natural and uncontrollable—weather. The dominant cause for differences between targets programmed and targets read out in the entire GAMBIT-3 program was cloud cover. In later years, the output of weather satellites lessened that effect, but it would persist as long as cloud cover data were other than instantaneous." Perry, op. cit., pp. 233-3.
71. PET Report, Mission 4301/66.
72. Ibid.
73. Ibid.
74. Ibid.
75. Perry, op. cit., pp. 232-3.
76. Ibid., p. 51.

References (continued)

77. Ibid., p. 257.
78. Ibid., p. 274.
79. Interview, Col. Lee Roberts (USAF-Ret), Jun 87.
80. Perry, op. cit., p. 293.
81. EKC "K" Orientation Briefing, 7 Mar 74; Perry, op. cit., p. 266.
82. PET Report, Mission 4332, 1971.
83. Perry, op. cit., pp. 297-98.
84. Memo, BGen D.D. Bradburn, Dir/SP, to Dr. J. McLucas, DNRO, 20 Mar 73, Subj. GAMBIT Mission Summary (4337).
85. Conversation with and data from Thomas Diosy, EKC, Jun 87; EKC "K" Orientation Briefing, op. cit.; Perry, op. cit.
86. Report "The Application of Image Forming Satellite Reconnaissance to Crisis Management," COMOR, 24 Jan 66.
87. Perry, Draft NRO History, Chap XVII, BYE 15649-75, p. 39.
88. Memo, D. Packard, DepSecDef, to R. Helms, DCI, L.A. DuBridge, PresSciAdvsr, and J.L. McLucas, DNRO, 16 May 69, subj: "Real-Time Readout."
89. Report "Requirements for Image Forming Satellite Reconnaissance Responsive to Warning/Indications Needs," prepared by COMIREX, 5 Jan 68.
90. Memo, A.H. Flax, DNRO, to Chm, USIB, 12 Mar 69, subj: "Study of Requirements for Image-Forming Satellite Reconnaissance Responsive to Warning/Indications Needs," in NRO policy files.
91. Minutes of NRP ExCom Mtg, 15 Aug 69.
92. Memo, E.H. Land, et al, to Dr. L.A. DuBridge, PresSciAdvsr, 12 Aug 69, no subj., in Land Panel papers, DNRO files.
93. Memo, G.T. Tucker, AsstSecDef (SA) to DepSecDef, 14 Feb 70 Subj: "Interim Report of the Committee for Immediate Recovery of Imagery (Fubini Committee)"; Report of the Committee for Immediate Recovery of Imagery, 16 Feb 70.
94. Memo, E.H. Land, et al, to Dr. L.A. DuBridge, 13 Mar 70, Subj: "Solid State Real-Time Readout System."
95. Memo E.H. Land, et al, to Dr. L.A. DuBridge, PresSciAdvsr, 13 Jul 70, Subj: "Photographic Reconnaissance Systems Status."
96. Director's Report to the NRP Executive Committee on FY-70 Status and FY-71 Program, 15 Jul 70.
97. Memo, J.L. McLucas, DNRO, to Dir/CIA Recce Prog, 27 Jul 70, Subj: "Approval of Electro-Optical Imaging Program System Definition Phase."
98. Report "Study of Intelligence Requirements for Crisis Response Satellite Imaging" COMIREX, Apr 71.
99. Perry, op. cit., manuscript Chap. XVII, pp. 90-91.
100. Senator Allen J. Ellender as Chairman of the Senate Appropriations Committee took a position on the issue which could not be ignored: Choose one of the two systems . . . we will approve funding for only one development—not both.
101. Memo, H.A. Kissinger to SecDef., Dir/OMB, DCI, PresSciAdvsr, Chm. PFIAB, 23 Sep 71, Subj: "Near-Real-Time Satellite Reconnaissance System."
102. Interview, Col. Lee Roberts, Jun 87.
103. Written Communication, MGen David D. Bradburn, (USAF-Ret), 16 Mar 67.
104. Perry, op. cit., pp. 304-14.
105. GAMBIT Program Preliminary Flight Evaluation Report for Flight No. 47, Dec 76.
106. GAMBIT Program Preliminary Flight Evaluation Report for Flight No. 48, p. 135, Jun 77.
107. Interview, Col. Lee Roberts (USAF-Ret), Jun 87.
108. Interview, Col. ▨▨▨▨ (USAF-Ret), Jun 87.
109. Interview, Col ▨▨▨▨ USAF-Ret), Jun 87.
110. Interview, Col. ▨▨▨▨ op. cit.
111. Letter from Col. ▨▨▨▨ SP-2 to SP-1 (B/Gen John Martin) "Analysis of GAMBIT Project," 24 Aug 67.
112. Ibid.

References (continued)

113. Data furnished by Capt. ███████████, Sep 87.
114. Refer to the best resolution of the mission. Average values generally were twice "the best," or more.
115. "Summary Analysis of Program 206 (GAMBIT)," incl to ltr. to DNRO Dr. Flax from SAFSP, B/Gen John Martin, 29 Aug 67.
116. Letter Hq, ARDC to Commander, AFCRC, Subject, "Request for Information," 9 Dec 55.
117. ███████ op. cit.
118. When General Power, arriving from SAC to head the ARDC, first heard these expressions, he observed that they were synonyms for "reading newspapers and drinking coffee." He described, with vigor, what would have happened, in SAC, to anyone found "goofing off" in these categories. Until his departure, their usage at Hq, ARDC was muted.
119. Ltr., BGen B.A. Schriever, Commander WDD, to Commander, ARDC, 30 Mar 55.
120. Later this list was formally requested by Col. ███████████ and appeared in Memo, subject, "Program Management," Battle to ███████ 5 Sep 61. The italicized sentences are by Battle.
121. Eisenhower, op. cit., p. 520.
122. Ibid, p. 522.
123. Lundahl interview, 12 Nov 86.
124. Robert F. Kennedy, *Thirteen Days, a Memoir of the Cuban Missiles Crisis*, Norton Company Inc., New York, 1969, pp. 23-24.
125. Lundahl interview, op. cit.
126. Ibid.
127. Ibid.
128. Director of Central Intelligence Directive No. 2/7, effective 9 Aug 60.
129. Director of Central Intelligence Directive No. 1/13, effective 1 Jul 67.
130. Letter, Subj, "KH-8 Mission 4354 System Operation Performance," dated 27 Aug 84.
131. COMIREX Automated Management System Functional Overview, TCS-5240-77, Jun 77.
132. Ibid.
133. NPIC Message, CITE ███████████ dated 4 Sep 81.
134. Contributed by Donald E. Welzenbach, CIA History Staff.

~~SECRET~~
~~NOFORN-ORCON~~

GAMBIT Index

AAA

A-12   31, 116,
Abalakovo   154
Ablative Shield   26
Ad Hoc Requirements Committee   16, 119
Advance Development Section   55
Advanced Reconnaissance System   17
Advanced Research Projects Agency (ARPA)   12
Aerial Recovery   4, 26, 29
Aeronautical Systems Division   49
Aerospace Corporation   20, 21, 29, 81, 103
Agena   6, 9, 13, 21, 37, 39, 49, 55, 58–61, 64–66, illus. 67, 72, 77–79, 94, 96, 103, 104, 180
 Agena-B   106, 180
 Agena-D   26, illust. 28, 59, 64
Agency for International Development (AID)   120
Aircraft Carrier, Soviet   156
Air Development Center (ADC)   107
Air Force Ballistic Missile Committee   7, 110
Air Force Ballistic Missile Division (AFBMD)   9, 10, 12, 20, 36, 110–112, 114, 177
Air Force Missile Test Center   110
Air Force Satellite Control Facility—see Satellite Control Facility
Air Force Special Weapons Center   107
Air Force Systems Command   20, 23, 36, 103
Air Force Weapons Board   19
Air Materiel Command   19
Air Research and Development Command   8, 10, 12, 20, 107–110
Air Staff   10, 12, 17, 19, 23
Alaska   4
Allen, Lew, Jr.   46, 79, 85, 96
Anderson, David   iii
Antarctica   43, 72
Anti-Ballistic Missile (ABM)   143, 179
Apollo-11   143
▬   168
ARGON, Project   179
Armed Forces Procurement Regulations   19
Army Air Corps   49
Army Map Service   179
Astronautics Directorate   10
Astro-Position Terrain Camera (APTC)   63, 64, 69, 76
Atlas   3, 6, 7, 9, 13, 23, 26, illust. 28, 45, 52, 55, 59–61, 105
Attitude-Control Subsystem   43, 59, 72, 97, 98
Austria   1, 2
AWACS Aircraft   159

BBB

B-52   2, 179
B-58   179
Back-Up Stabilization System (BUSS)   38, 59, 64, 72
Balkans   1
Ballistic Missiles   3
Ballistic Missile Division—see Air Force Ballistic Missile Division (AFBMD)
Balloon Reconnaissance   4, 5
Baltic Sea   1
▬   33
Battle, Lee   110–113
BEAR-F Aircraft   138
Beck, Frank   117
Belden, Thomas G.   50, 108
Belden Study   108
Bell Telephone Laboratories   40
Bell Engine   81
Belorussian Military District   153
Berezniki Chemical Combine   173
Berg, Russell   37, 46
Berlin   2
Bi-Mat Film Processing   86, 87
BISON Aircraft   2
Bissell, Richard M., Jr.   9, 10, 112, 177
Black Contracting   20, 39, 177
BLACKJACK Bomber   162
Blanket   15, 17
Block-II   75, 77–79
Block-III   81
Block-IV   84
Bombs-in-Orbit   23, 35
Boston, MA   181
Box-Level Testing   32
Bradburn, David D.   iii, 37, 84, 95
Bulganin, Premier   4
Bushmann, Rudi   iii
Buzard, Frank   113
BYEMAN Security System   14, 122, 123

CCC

C-119   29, 41
C-130   illust. 77
CSA-1 Missile   131
Cabell, Charles P.   16
▬   170
California   107, 181

-187-

~~SECRET~~

Cambridge Research Center, MA   107, 108
Camera-Carrying Balloons   4
Camouflage, Concealment, and Deception (CC&D)   166
Canada   29
Caspian Sea Monster   164, 165
Cassegrain Telescope   181
Catadioptric Lens   15, 57, 62
CBS Laboratories   87
████████   168
Central Intelligence Agency (CIA)   iii, 2, 5, 8, 9, 11, 13–15, 87, 88, 91, 103, 105, 110, 112, 115, 119, 120, 128, 129, 177, 179
Cer-vit Glass   58
Charyk, Joseph V.   iii, 11, photo 12, 15, 17–21, 23, 24, 29, 30, 33–40, 44, 179
Chestnut Street Facility   18
████████   iii
Chief of Naval Operations   77
China   3, 138, 143, 148, ████
████████
Churchill, Winston   1, 2
Cis-Lunar Defense/Space   10
Civil Service Commission   109
Civil Applications Committee (CAC)   120
████████   iii
Colby, William   122
Cold-Gas System   25, 26, 38
Color Film/Photography   41, 85, 100, 173
Columbia University   2
Commander-In-Chief, Pacific (CINCPAC)   77
Committee on Imagery Requirements and Exploitation (COMIREX)   iii, 87, 88, 91, 99, 120–122, 128, 130
COMIREX Requirements Structure (CRS)   124, 126
COMIREX Automated Management System (CAMS)   126, illust. 127
Committee on Overhead Reconnaissance (COMOR)   16, 17, 41, 71, 86, 87, 119, 123, 124
Communications Satellites   23
Comsat Corporation   40
Contracting Warrant   19
████████
Convair   50
Convergent-Stereo Camera   15
CORONA   iii, 5, 6, 8–12, 14–16, 18, 20, 22, 26, 29, 30, 32, 34, 37–39, 46, 47, 53, 56, 61, 72, 75, 81, 86, 97, 107, 112, 113, 116, 119, 120, 123, 126, 136, 177–181
CORONA H-30 Recovery Vehicle   30
████████   30, 58
Cosmos-264   75, photo 76

Crateology   138
████████   photo 99, 100
Cristy, J.   65
████████ Program   101
Cuban Missile Crisis   34, 86, 116
Cue Ball Program   22–24, 35
Czechoslovakia   2, 86, ████████

DDD

Dayton, OH   49
Davidson, H.   65
Deep-Space Radar Facility   158
Defense Intelligence Agency (DIA)   74, 91, 119, 128
Defense Mapping Agency   175
Delta-Class Submarine   157
Department of Agriculture   120
Department of Commerce   120
Department of Defense   13, 50, 77, 122, 128, 129
Department of Interior   120
Department of State   119, 128
Design & Analysis Division   91
Delta Dagger   50
Diodes—see Light-Sensing Diodes
Diosy, Tom   iii
Director of Central Intelligence (DCI)   16, 122, 123, 179–181
Director of Central Intelligence Directive (DCID) No. 2/7   16
Director of NRO   40, 56, 85, 105, 121, 179–181
Directorate of Science & Technology   88, 91
Dirks, Leslie   87, photo 91
Discoverer, Project   5, 6, 9, 110–113, 177
DoD Directive 5200.13   12, 135
DOG HOUSE Radar   138
████████ Project   88, 90, 91
Double-Recovery Vehicle   76
Douglas Aircraft Co.   6, 177, 180
Dual-Mode GAMBIT   97, 98, 103
Dual-Platen Camera   64, 84, 85, 100, 101, 103, illust. 161, 173
Dual-Recovery Vehicle   75, 76, 78
DuBridge, Lee   89, 90, photo 91
Duckett, Carl   88, photo 89
Dulles, Allen W.   16, ████████   46

EEE

E-1 Program   17, 86, 87
E-2 Program   17, 86, 87
E-3 Program   18, 86
E-4 Program   18

E-5 Program   18, 179–181
E-6 Program   17–22, 30, 34
East Germany   153
Eastern Asia   2
Eastern Europe   2
Eastman Kodak Company (EKC)   iii, 15, 17–22, 25, 30–33, 36, 49, 55–61, 64, 66, 67, 72, 79, 81, 84, 85, 93, 96, 105, 106, 112
Edwards AFB, CA   50, 107
███████   30, illust. 31, 58
Eglin AFB, FL   107
Egypt   86, ███████   121
Eisenhower, Dwight D.   1–5, photo 5, 9, 10, 15, 17, 106, 110, 115–117, 177
Electronic Intelligence (ELINT)   179
Electro-Optical Imaging System (EOI)   86–89, 91–93, 101, 103
Electrostatic-Tape Camera   87
Environmental Protection Agency (EPA)   120
Estonia   1
███████   168
Eurasian Landmass   132
Europe   1, 2
Exemplar, Project   22, 23
Exploitation Subcommittee (ExSubcom)   120

FFF

Factory-to-Pad Concept   56, 72, illust. 73
FAGOT Aircraft   131
███████   168
False-Color Infrared Film   85
Farnborough, England   81
Fascist Italy   1
Feak, D.   65
FEEDBACK, Project   6
Film-Read-Out GAMBIT (FROG)   86–93
Film-Recovery System   18, 36
Finland   1, 2, 182
FISHBED-J Aircraft   131, 132
Fitzpatrick, James C.   iii
FLAGON Aircraft   131
Flax, Alexander   33, photo 44, 46, 71, 76, 85, 88
Florida   107
Flying-Spot Scanner   17
Ford, Ralph J.   46
Foster, F.G.   88
Foster, John   88
France   1
Freon Gas   25
Fubini, Eugene   88, photo 89, 90
Funk, Ben I.   36
Fused-Silica Mirror   58

GGG

GAMBIT-Cubed   59
GAMBIT Programmer   45, 47, 48
GAMBIT Program Commendation   117
GAMBIT-1   26, 27, 42–44, 53, 58, 59, 65–67, 70–72, 94, 103, 104, 116, 123
GAMBIT-2   57
GAMBIT-3   57–81, 84–86, 94–98, 101, 103–105, 116
GAMBIT-4   57
Geary, Leo   46
General Electric Co. (GE)   18–21, 25, 29, 30, 32–38, 45, 47–49, 51–53, 55, 58–60, 64–67, 72, 93, 105, 106, 112, 177, 182
GENETRIX, Project   4, 6
Geneva, Switzerland   115, 117
Geneva Summit Conference   3
Germany   1
  East   153
  West   2
Gillette, Hyde   7
Gillette Procedures   7, 8, 10, 12, 36
Global Weather Central (GWC)   132
Golden-Finger Award   93
Golden Gate Bridge   163
Gorman, Frank B.   iii, 55
Greer, Robert E.   17–24, photo 19, 29, 30, 32, 34–39, 41, 44, 45, 50, 51, 56, 58, 59, 64, 65, 72, 112, 113
GRIFFON (SA-5) Missile   179
Guatemala   2
Guided Missile Secretariat   7
GUIDELINE (SA-2) Missile   179
Gulf of Finland   182

HHH

Haig, Thomas   37
███████   iii
Harley, John   64, 65
Harvard University   181
Hawaii   96, 99
███████   91, 92
Helms, Richard M.   71, 88, photo 89, 117
HEN HOUSE Radar   138
Heran, Paul J.   photo 18, 23, 37, 46
HEXAGON, Project   70, 97, 98, 101, 139, 141, 171, 172
Hicks, Frank   32
High-Density Acid (HDA)   78, 97
Highboy   97
Higherboy   97
Hill, Jimmie D.   iii

Hitch-Up 37, 39
Hitler, Adolf 1
Holloman Air Development Center, NM 107
Horizon Sensor 33, 98
Huntley, Harold 64, 65
HYAC (High-Acuity) Camera 6
Hydrazine 96, 97
Hydrogen Bomb 2
Hypergolic Propellants 25, 26

III

ICBM 16, 52, 124, 126, 143, 148, 153, 155, 168
IRBM 143
IR Film—see Infra-Red Film
Image-Motion Compensation 25, 27, 39
Imagery Collection Requirements Subcommittee (ICRS) iii, 99, 120
Incentive Contract Structure 51, 53, 72
Infra-Red Film 100, illust. 166, 167, 173
Inglewood, CA 7, 17, 177
Inlow, Roland iii, photo 91, 120
Integral Secondary Propulsion System (ISPS) 98
Intelligence Community iii, 6, 13, 14, 16, 42, 51–53, 55, 71, 75, 86, 87, 91, 95, 99, 115, 119–123, 126, 153, 177, 181
Intelligence Community Staff 96
████████ 31, 57, 58
Iran 2
Iron Curtain 2
Israel 86, ████
Italy 1, 2
Itek Corporation 179–181

JJJ

Jacobson, Ralph 93
Johnson, Clarence L. "Kelly" 4
Johnson, Lyndon B. 129
Johnson, William 113

KKK

Kaz-A Aircraft 162
KH-1 Camera 177
KH-2 Camera 177, 179
KH-3 Camera 177
KH-4 Camera 14, 97, 116, 177, 181
KH-5 Camera 179
KH-6 Camera 37, 179–182
KH-7 Camera 25, 103, 104, 116, 122, 123, 168, 182
KH-8 Camera 69, 103, 104, 116, 121, 122, 141, 168
KH-9 Camera 70, 97, 139, 141
████████ 88, 90, 91
KH-11 Camera 93, 101
Kazan Airframe Plant 162
Kennedy, John F. 15, 116, 129
Kennedy, Robert F. 115, 116
████ Project 93, 101
Khrushchev, Nikita 4, 115, 153
Kierton, R. 65
Killer Satellite 76
Killian, James R., Jr. 1, photo 5, 9, 10
King, William G. iii, 36–39, 41, 44–49, 52, 53, 58, 59, 64, 70, 72, 74, 75, 79, 112, 113
Kirtland AFB, NM 107
Kissinger, Henry 92
Kistiakowsky, George B. 9–12, photo 11
Koche, Robert 94
Komsomolsk Shipyard 140
Korea 2
████████ 94, 95
Kueper, Robert 64, 65

LLL

Land, Edwin H. photo 11, 15, 86–89, 92, 181
Land-Recovery System 29, 30
LANYARD, Project 37, 38, 58, 179–182
Laser-Scan System 90
Latvia 1
Leningrad 179–182
Leverton, Walter F. 81
Lifeboat 37–39, 41, 42, 64, 72, 113
Light-Sensing Diodes 86
Lincoln Labs 10
Lincoln Plant EKC 22
Lithuania 1
Lockheed Corp. 4, 20, 21, 38, 47–49, 55, 57, 59–61, 64–67, 72, 94, 96, 97
Lockheed Missiles & Space Co., Inc. (LMSC) iii, 38, 47, 70, 93, 100, 105, 106, 177, 180, 181
London, England 81
Los Angeles, CA 12, 47, 48, 94, 112, 121
Lowest-Level-of-Assembly Testing 33
Lunar Reconnaissance/Landing Program 94
Lunar Orbiter/Surveyor 66, 94
Lundahl, Arthur iii, 115–117, photo 129

MMM

MacLeish, Kenneth 32
Mahar, James 32, 56
Maksutov Lens 25, 57
Manned Orbiting Laboratory (MOL) 88, 91
Martin, John L., Jr. iii, 11, photo 12, 29, 33, 44–49, 51–53, 70, 72, 75, 79, 112, 113
Martin-Marietta Co. 65, 93, 106
Martin Motivator 53
Massachusetts 107
Massachusetts Institute of Technology (MIT) 1, 10
███████ 16
███████ iii, photo 99, 100
McCone, John A. 179, 181
McElroy, Neil 16
McLucas, John L. 77, photo 84, 85, 87, 88, 91, 96
McMillan, Brockway photo 39, 44, 56, 58–61, 64, 65, 85, 182
McNamara, Robert S. 179
Meeting the Threat of Surprise Attack 5
Mexico 29
Midas Project 23
Middle East Crises 78
Missile Assembly Building (MAB) 49
Mitchell, Leslie iii, 32
Mitre Corporation 85
Mono-Cubic-Dispersed Film 85
Moore, Richard 113
Murphy's Law 93
MURAL Camera 181
Myacheslav M-4 Bomber 2

NNN

NATO Designator 179
National Aeronautics and Space Administration (NASA) 23, 61, 66, 95, 96, 105, 171
National Imagery Interpretation Rating Scale (NIIRS) 130, 173
National Intelligence Estimate (NIE) 153
National Photographic Interpretation Center (NPIC) iii, 13, 94, 96, 115, 119, 128, 129, 132
National Reconnaissance Office (NRO) iii, 6, 13, 14, 20, 33, 44, 46, 448, 55, 57, 58, 61, 66, 70, 77, 88, 89, 94, 105, 112, 117, 119, 120, 121, 123, 126, 130, 132, 136, 171, 182
National Reconnaissance Program (NRP) iii, 13, 14, 70, 123
National Reconnaissance Program Executive Committee (ExCom) 71, 74, 87, 88, 90–92
National Science Foundation 10

National Security Adviser 92
National Security Agency (NSA) 13, 119
National Security Council (NSC) 9, 17, 21, 33
National Security Council Intelligence Directive (NSCID) No. 8 128
National Tasking Plan 128
Nazi Germany 1
New Mexico 107
New York 107
Nikolayev Shipyard 156
Nixon, Richard M. photo 92, 122, 129
NRO Program A 46, 87, 88, 93, 99, 112, 181
NRO Program B 46, 87, 88, 91, 181
NRO Program C 46
NRO Program D 46
North Atlantic Treaty Organization (NATO) 2
North Korea 76, 116
Norway 4
Nuclear Weapons 2
"Null" Program 22–24

OOO

Oahu, HI 13, 41, 177
Ocean-Recovery Program 29
Oder, Frederic C.E. iii, 32, 70
Office of Management and Budget (OMB) 91
Office of Missiles and Space Systems 12
Office of Space Systems 13
O'Green, Fred 47
Omaha, NE 132
On-Time Delivery 108, 109, 113
"Open Skies" 3, 4, 115, 117
Operating Division-4 (OD-4) 121
Optical-Figure Error Budget 31
Orbit-Adjust Module 59
Orbital-Control Vehicle (OCV) 19, illust. 24, 25, 26, 30–33, 35, 37–39, 41, 42, 45, 47, 52, 53, 55, 56, 58–60, 72, 180, 182
Overhage, Carl 10
Overhead Reconnaissance 5
███████ 58
OXCART, Project 15, 31, 33, 116

PPP

P-Camera 181
Pacific Missile Range 23
Pacific Ocean 45, 77, 81
Pacific Recovery Area 30
Packard, David 87, photo 88, 89
Page, Hilliard 29
Pakistan 4

Panoramic Camera 14
Patrick AFB, FL 107, 110
Pavlograd 138
Payload-Adapter Section 56
Pearl Harbor, HI 2
Pentagon 12, 19, 44, 85, 120
Performance Evaluation Team (PET) 69
Perry, Robert iii, 24
Phased-Array Radar 154
Philadelphia, PA 18, 45, 47, 48, 52, 53, 72, 112
Photographic Intelligence (PHOTINT) 120
Photographic Payload Section (PPS) 59, 60, 65–67, 70, 75, 84, 93, 98, 106
Photographic Satellite Vehicle 69, 71
Photo-Sensitive Electrostatic Tape Camera 18
Photo Working Group (PWG) 120, 124
Pied Piper Project 6, 8
Pietz, John 32
Pioneer Satellite Program 17
Plesetsk Missile Test Range 155
Plummer, James W. 64, 65, 70
Poland 1
Polaroid Corp. 15
Powell, Robert M. iii, 64, 65, 70
Power, Thomas S. 20, 107
Powers, Francis Gary 4, 16
President's Board of Consultants on Foreign Intelligence Activities (PBCFIA) 10
President's Foreign Intelligence Advisory Board (PFIAB) 33
Presidential Reserve Funds 20
Presidential Science Adviser 1, 9, 90, 91, 110
Programs A, B, C, and D 46, 87, 88, 91, 93, 99, 112, 181
Program 206 32, 35, 36
Program 307 22, 23
Program 417, Weather Satellites illust. 133, 136
Programmed Integrated Acceptance Test (PIAT) 67
Purcell, Edward M. 181
███████ 58

RRR

R-5 Lens 78, 79
R-361 Optical Design 79
Ragusa, Peter iii, 70, 96
Raincoat 11, 18, 35
Ramo-Wooldridge Company 110
Rand Corp. 4, 6, 7
Readout Photography 9, 10, 17, 36, 72, 86–93
Reagan, Ronald 117

Reber, James Q. 16, 46, 119
Reconnaissance Laboratory (Wright Field) 15, 17
Recovery/Reentry Vehicle (RV) 6, 18, 19, 25, 26, 29, 34, 36, 41, 60, 64, 72, 106
Red Army 1, 2
Red China 2
Redundant Attitude-Control System (RACS) 75
Ricks, LtCol 46
Riepe, Quentin A. 23, 29, 30, 34, 36, photo 39
Ritland, Osmond J. 9, 12, 36, 49, 177
Roberts, Lee iii, photo 79, 93, 94, 97, 99
Rochester, NY 22, 32
Roll-Joint 37–39, 42, 55, 57–59, 61, 64, 66, 67, 70, 72, 75, 81, 113, 123, 180, 182
███████ 138, 168
Rome Air Development Center, NY 107
Rostow, Walter W. 3
Royal Aircraft Establishment 81
Ruebel, John W. 34
Russian Empire 1
Ryan-147 Drone 116

SSS

SA-1 Missile 131
SA-2 (GUIDELINE) Missile 131, 132, 179
SA-4 Missile 132
SA-5 (GRIFFON) Missile 138, 179, 182
███████ 25
SLBN 160
SS-9 Missile 138
SS-11 Missile 124
SSBN 139
SAFSP—Secretary of the Air Force Special Programs iii, 7, 17, 21, 23, 24, 33–37, 44, 46, 50, 55–61, 66, 67, 72, 75, 77–79, 87, 95, 99, 100, 105, 112, 120
SAFSP-6 55
SAFSS—Secretary of the Air Force Space Systems iii
Samos Project 6, 8–13, 16, 17, 19–23, 30, 32, 34–36, 86, 179
Samos Working Group 19
San Cristobal, Cuba 116
San Francisco, CA 163
Sary Shagan Missile Test Center 179
Satellite-Control Facility 20, 74, 77, 78, 113, 181
Satellite-Control Section (SCS) 59, 60, 65, 67, 69, 84, 93, 97–99, 105
Satellite Intelligence Requirements Committee (SIRC) 16, 119
Satellite Interceptor 23
Satellite Operations Center (SOC) 120, 121

Satellite Reconnaissance 4
Satellite Recovery Vehicle (SRV) 21, 59, 98, 181
Satellite Test Center 45, 79, 121
███████████████████████ 81, 95, 100, 171, 172
███████████████ 75, 78, photo 79
Schriever, Bernard A. 8, 10, 20, 23, 36, 110
Schultz, George 92
Scoville, Herbert, Jr. 179
SCUD Missile 169, 170
Search Mode iii, 124
Seay, James 18, 46
Secretary of Defense 7, 9, 13, 16, 88, 92, 122, 179
Seiman Stars 138
Senate 92
Sentry Project 6, 8
Severodvinsk Shipyard 157, 160
Sewell, John M. 32
███████████████ iii, 120
Sharp, Dudley C. 19
Sheldon, Huntington D. 46
Shuangchungtzu Missile Test Center 147
Signals Intelligence (SIGINT) 90, 100, 120
SIGINT Overhead Reconnaissance Subcommittee (SORS) 89
Silo Hardness 155
Simmons, Arthur B. 15, 17, 32, 49
███████ 169
Six-Day War 86
Skylab 81, 94–96, 171
███████ 46
███████ 74, 75
Somalia 168
South Africa 81
Southeast Asia 77, 179
South Pole 42
Soviet Aircraft Carrier 156
Soviet Ballistic Missile Submarine (SSBN) 139
Soviet-China Border 74, 138
Soviet Union 1–4, 6, 10, 16, 21, 33, 41, 52, 74, 100, 115, 117, 132, 136, 138, 143, 148, 153–155, 164, 169, 171, 182
Soyuz Spacecraft 171, 172
Space Systems Division (SSD) 12, 13, 17, 20, 23, 34–36
Space-Ground Link Subsystem (SGLS) 75
Special Assistant for Reconnaissance 13
Special Projects Group EKC 15
Spin Scan 87
Spoelhof, Charles P. 56
Spotting Camera 14
Sputnik 8, 10
SR-71 116
Standard Agena 64

START Talks 155
State Department—see Department of State
Stellar-Index Camera 69
Stevens, Don 32
Steward, James T. 46
Strategic Air Command (SAC) 7, 19, 20, 108, 132
Strategic Arms Limitation Talks (SALT) 88, 143, 148
Suez Canal 78
███████ 120
Sunnyvale, CA iii, 13, 45, 70, 113, 121, 177, 181
Sunset Strip 15, 17, 18, 20
Supplementary Solar Array 97
Surface-to-Air Missile (SAM) 143
Surprise Attack 1, 2
Surveillance Mode iii, 14, 123
Sylvester, Arthur 11

TTT

T-54 Tank 130
T-55 Tank 130
T-62 Tank 74
Taiwan 4
TALENT-KEYHOLE Security System 122, 123
Tallinn, Estonia 179, 182
Tape-Storage Camera 88, 91
Taylor, Rufus 46
Technological Capabilities Panel 4, 181
Telemetry 42, 45, 75, 84, 181
Tennessee 107
Terrain Camera 69
Thermal-Control Subsystem 32
Thiokol Corp. 180
Thor Booster 6–9, 180
Thrust-Augmented Thor (TAT) 180, 181
Titan ICBM 6, 7, 9
Titan-III 55, 59–61, 65, 69, 74, 93, 96, 106
TRW Corp. 93
Tucker, Gardner 88, photo 89, 90
Tullahoma AFB, TN 107
Typhoon Submarine 160
Tyuratam Missile Test Center 115, 138, 143

UUU

U-2 4, 5, 10, 16, 21, 23, 116, 119, 179
Ultra-Thin-Base Film 74
Union of Soviet Socialist Republics (USSR)—see Soviet Union
US Army 2, 119, 128, 179
US Congress 49, 50, 92

US Intelligence Board (USIB)   13, 16, 41, 70, 71, 74, 86, 88, 94, 119
US Military Academy   1
US Navy   77, 119, 128
USS Pueblo   116
Ultra-Low-Expansion (ULE) Glass   58
Utica, NY   45, 47, 48, 60

VVV

Valley Forge, PA   32, 45
VALLEY Program   36, 56
Vandenberg AFB, CA   13, 20, 41, 45, 49, 57, 70, 94, 113, 177, 182
Vehicle Atmospheric Survivability Test (VAST)   78, 81
Venezuela   81
Vietnam   168

WWW

Waggershauser, Herman   15, 49
Walter Reed Army Hospital   115, 117
Warsaw Pact   138
Washington, D.C.   37, 92, 96, 110, 112, 128
Weapon System Project Office   107
Webb, Vernard   181
Welzenbach, Donald E.   iii
Wendover AFB, UT   29
West Coast Contract Office   179
West Germany   2
Western Development Division (WDD)   7, 8, 10, 110

Western Europe   2-4
West Point   1
White House   2, 20, 110, 112, 115, 116
White, Thomas D.   19
Wonsan Harbor, Korea   116
World War II   91
Worthington, Roy   113
Worthman, Paul E.   iii
Wright Air Development Center (WADC)   107, 108, 114
Wright Air Development Division   17
Wright Field   50, 107-109
WS-117L   6-8, 17, 177
WS-461L   6

XXX

X-Ray Examination   52

YYY

Y-Class Submarine   139, photo 140
Yeager, Charles   50
Yevpatorivo Radar Facility   158
███████   53
Yom Kippur War   138, 169
Yurya   173, 174

ZZZ

███████   87, 91-93
Zhukov, Georgi K.   1
███████   87

Center for the Study of National Reconnaissance Classics

GAMBIT PROGRAM
EASTMAN KODAK CO.
1977 PRESENTATION

CENTER FOR THE STUDY OF
NATIONAL RECONNAISSANCE
CHANTILLY, VA

APRIL 2012

TOP SECRET G

BIF008W-C-015446-OH-77
10-19-77
1 of 32 Pages
-001

GAMBIT PROGRAM

EASTMAN KODAK CO.

Handle via BYEMAN
Control System Only

TOP SECRET G

BIF008W-C-015446-OH-77

K PROGRAM

PROVIDES PHOTOGRAPHIC PAYLOAD FOR PROGRAM 110, THE GAMBIT DUAL RECOVERABLE SATELLITE RECONNAISSANCE SYSTEM.

PROGRAM 110 OBJECTIVE

ACQUIRE VERY HIGH RESOLUTION STEREO PHOTOGRAPHS OF SELECTED TARGET AREAS ANYWHERE ON EARTH.

Handle via BYEMAN
Control System Only

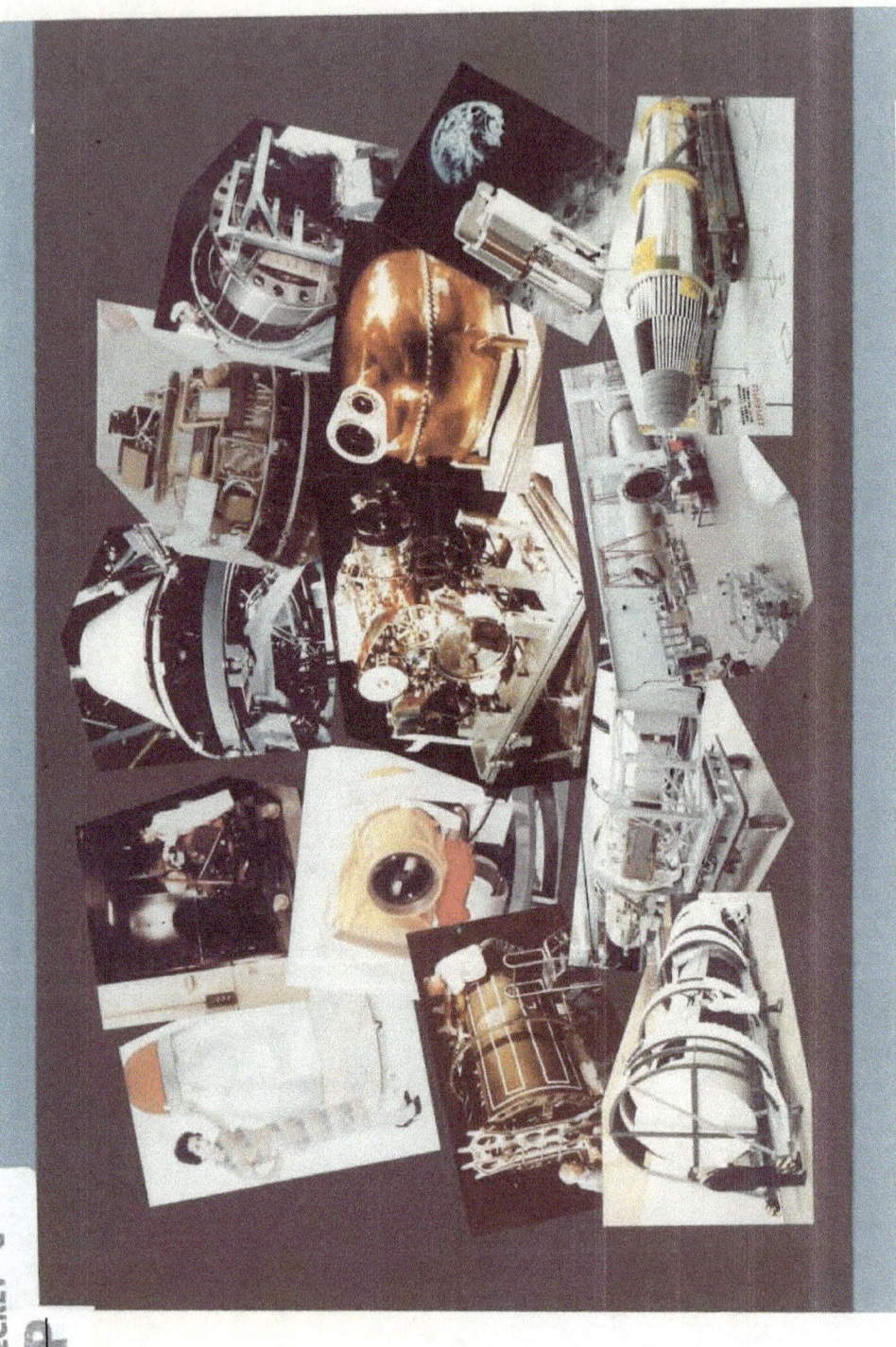

GAMBIT PROGRAM - Eastman Kodak Co. Presentation

BIF008W-C-015446-0H-77

~~TOP SECRET~~ G

Apr. 16

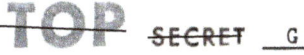

~~TOP SECRET~~ G

Handle via BYEMAN
Control System Only

Evolution of Gambit Systems

LAUNCH DATES

1963/66 GAMBIT PROGRAM
- 3 FT. DIA. VEHICLE
- 30+ FLIGHTS

1966/69 ADVANCED GAMBIT
- 5 FT. DIA. VEHICLE
- SINGLE RECOVERY BUCKET
- FLIGHTS 1-22

1969/76 ADVANCED GAMBIT (FO-11)
- DUAL RECOVERY BUCKETS
- FLIGHTS 23-47

1976/82 ADVANCED GAMBIT (FO-7)
- DUAL RECOVERY BUCKETS
- DUAL CAMERA SYSTEMS (9" & 5")
- FLIGHTS 48-54

1982-
- FLIGHTS 55-60

BIF008W-C-015446-OH-77

Program Milestones

PROGRAM START — 1964

FIRST LAUNCH	1966
ULTRA THIN BASE FILM	1967
USE OF SO-121 COLOR FILM	1968
DUAL SRV'S · 14 DAY ORBITAL LIFE	1969
USE OF 1414 HIGH DEF. B&W FILM	1970
LOW COEF. OPTICAL MATERIALS	1970
PPS FACTORY TO PAD	1970
20 DAY ORBITAL LIFE	1971
LENS FORMULA CHANGE R·5	1971
30 DAY ORBITAL LIFE	1972
EXPOSURE SLIT CHANGE	1972
INCREASED FILM CAPACITY (10,880)	1973
IMPROVED OPTICAL QUALITY	1973
USE OF SO-124 HIGH DEF. B&W FILM	1973
USE OF SO-131 FALSE COLOR IR FILM	1973
45 DAY ORBITAL LIFE	1974
9 × 5 DUAL PLATEN CAMERA	1977
75 DAY ORBITAL LIFE	1977

GAMBIT PROGRAMS

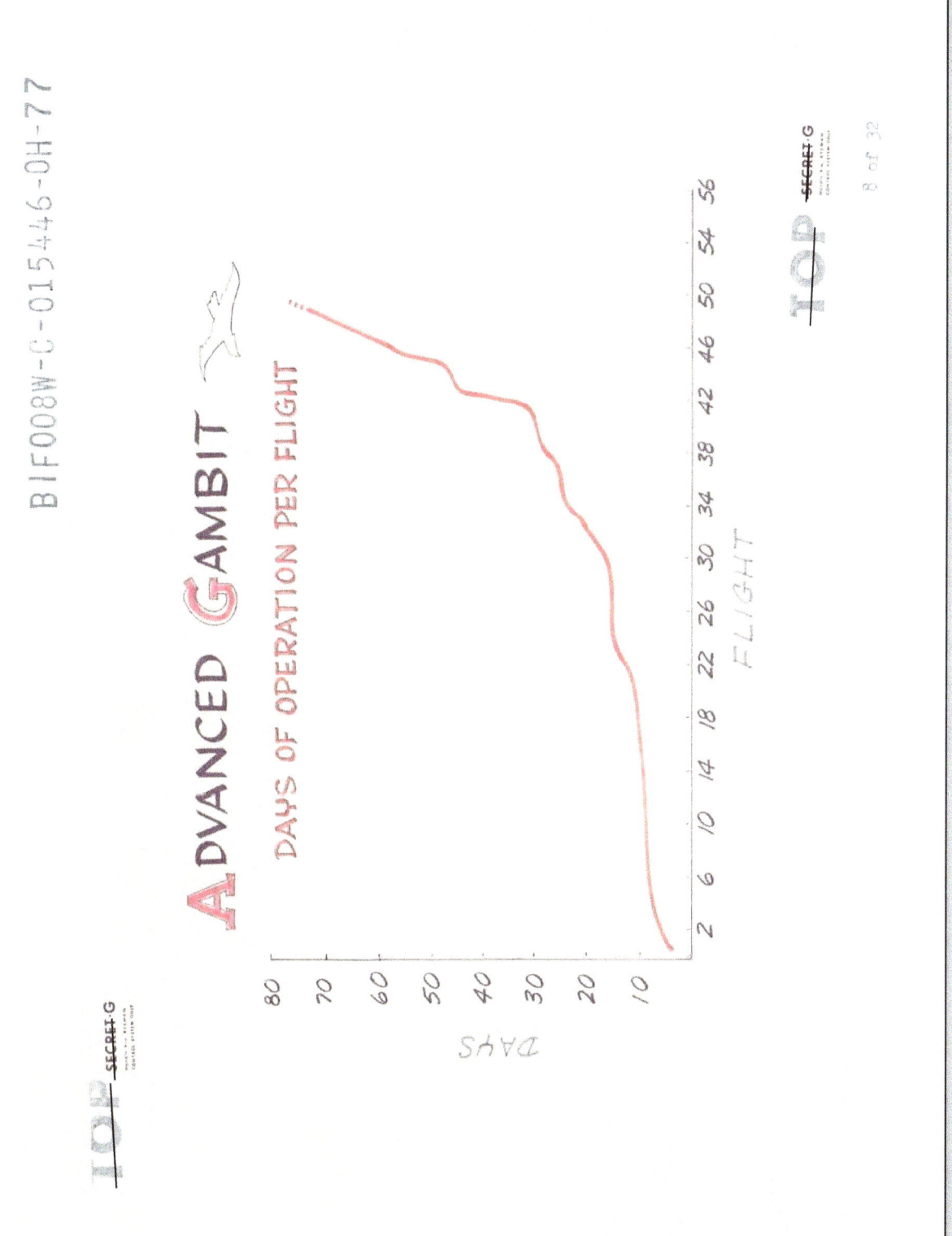

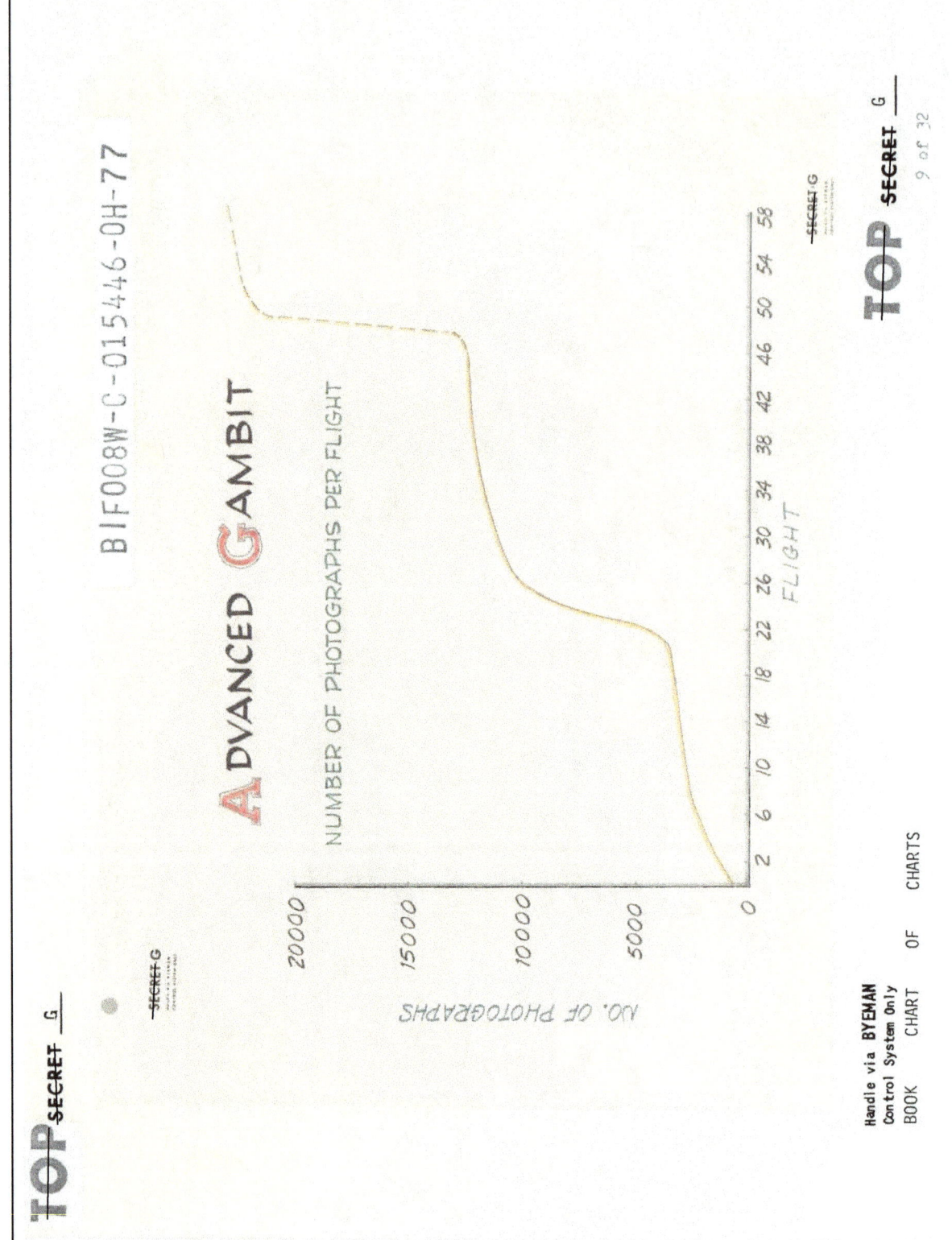

Advanced Gambit

– History of Ground Resolution

- IMPROVED FILMS
- DUAL GAMMA PROCESS
- 175" F.L.
- IMPROVED OPTICS

GROUND RESOLUTION

FLIGHTS: 2, 6, 10, 14, 18, 22, 26, 30, 34, 38, 42, 46, 50, 54, 58

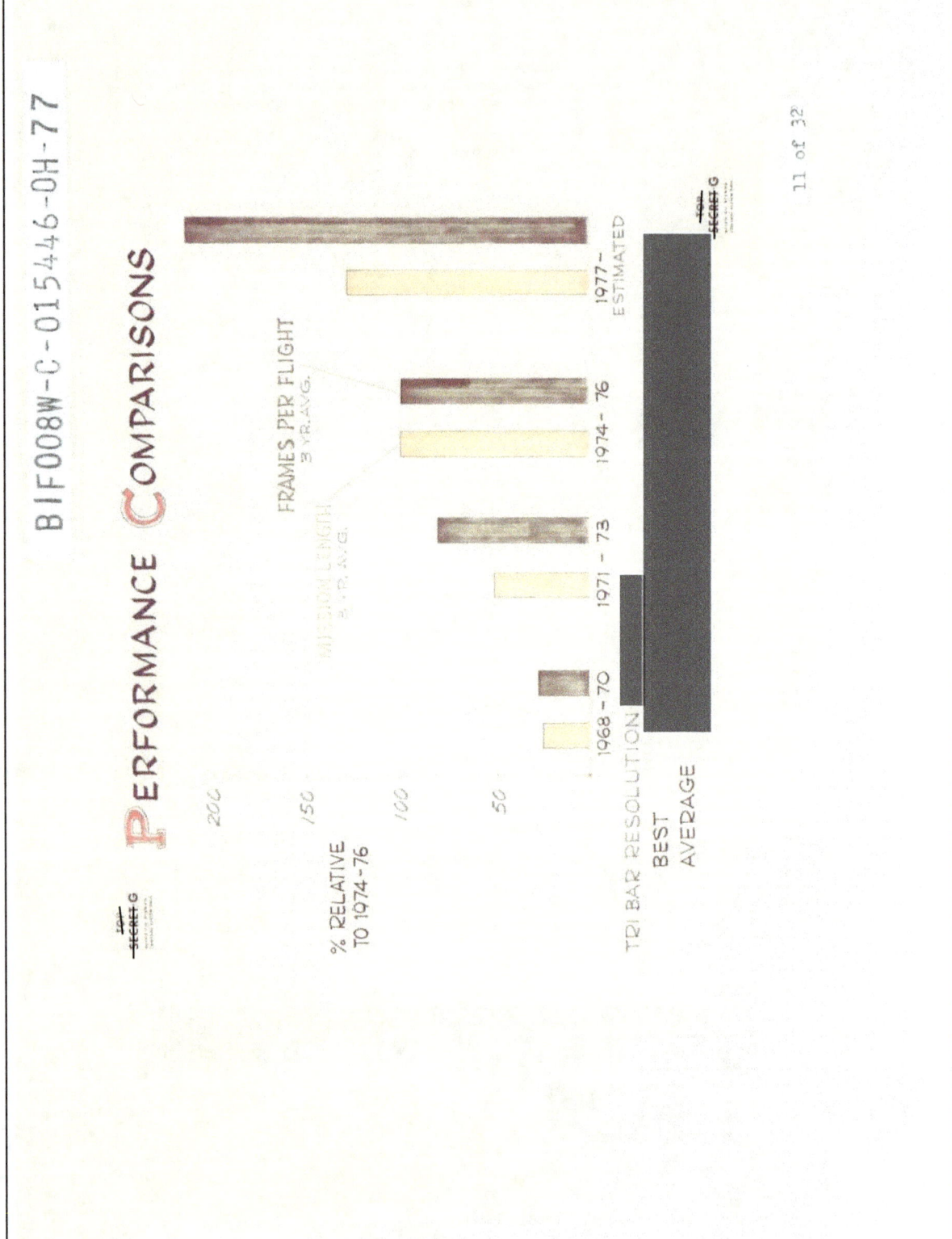

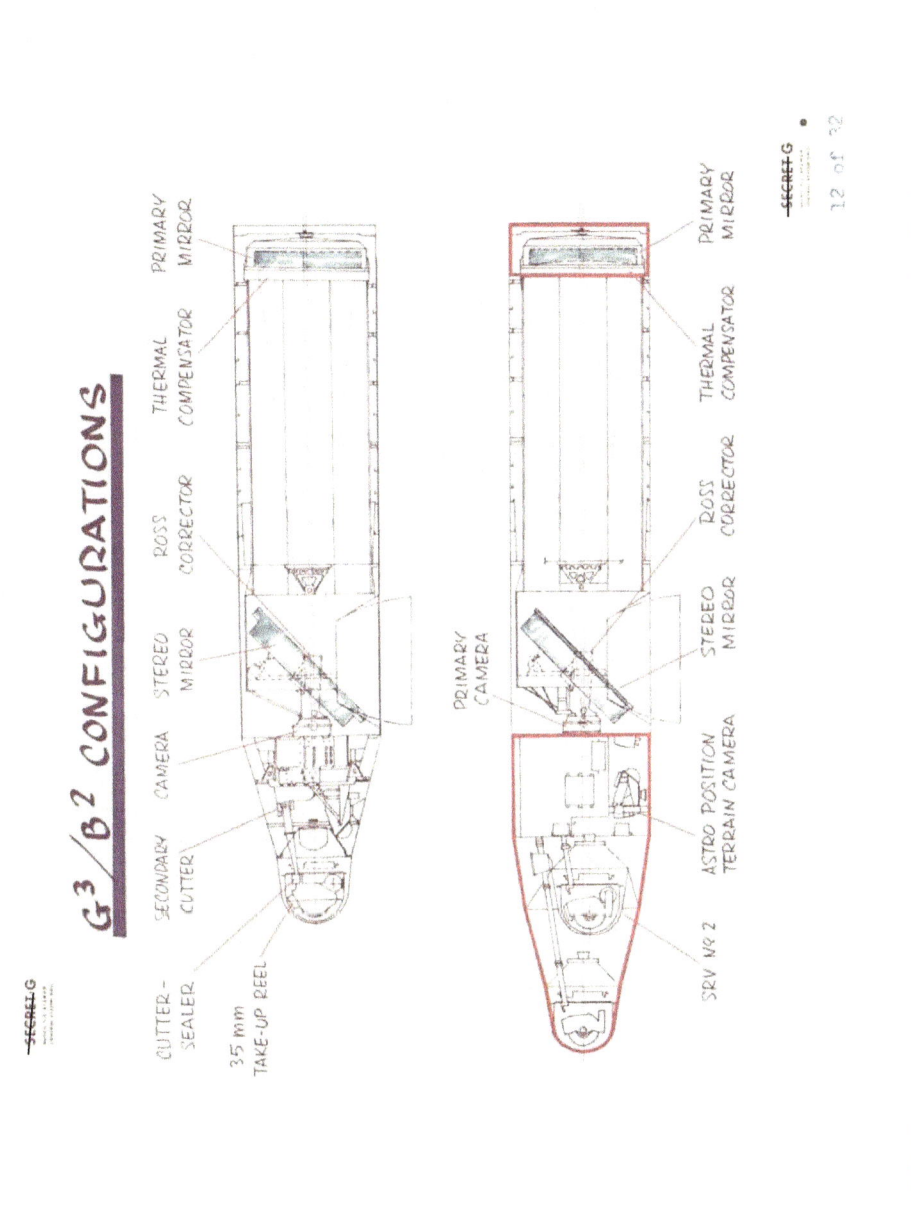

BIF008W-C-015446-OH-77

FRC GAMBIT PHOTOGRAPHIC PAYLOAD SYSTEM
FM-48-54 FLIGHT USE 1977-1980
INCORPORATE 9" & 5" CAMERA SYSTEMS

Labels:
- 9½" FILM SUPPLY ENCLOSURE
- 9½" FILM SUPPLY
- 9½" RAM
- 9½" TUNNEL SEAL & FILM TRAP
- 9½" FILM CHUTE
- 5" TAKE-UP
- 5" FILM CHUTE
- 5" TAKE-UP
- 5" TSRT
- DUAL PLATEN CAMERA
- 5" FILM SUPPLY ENCLOSURE
- 5" FILM SUPPLY
- 5" LOOPER
- 5" RAM
- LENGTH 28.5'
- DIA 60"

Handle via BYEMAN
Control System Only

BIF008W-C-015446-OH-77

Gambit Configuration

ON PAD

VEHICLE	LENGTH (ft.)	WEIGHT (lbs.)	ENGINE THRUST (lbs.)
TITAN III B			
STAGE 1	77.8	311,000	450,000
STAGE 2	23.3	74,000	102,000
PSV			
SCS	18.3	17,300	17,000
PPS/PAS	30.1	5,300	
TOTALS	149.5	407,600	

ON ORBIT

OVERALL LENGTH (ft.) 48.4
INJECTION WEIGHT (lbs.) 9,000
PROPULSIVE EXPENDABLES (lbs.) 300
TOTAL FILM WEIGHT (lbs.) 160
SATELLITE RECOVERY VEHICLE WEIGHT (lbs.) 400
RECOVERY VEHICLE WEIGHT AT AIR SNATCH (lbs.) 200

PPS
PSV ROLL JOINT
SCS
MABA
LAUNCH VEHICLE
TITAN IIIB

Handle via BYEMAN
Control System Only

ORBITING VEHICLE CONFIGURATION

- RV 2
- RV 1
- PRIMARY CAMERA
- ROSS CORRECTOR
- STEREO MIRROR
- ROLL JOINT
- FORWARD EQUIPMENT RACK
- PROPELLANT TANKS
- AFT RACK
- SECONDARY PROPELLANT SYSTEM
- FUEL TANK
- OXIDIZER TANK
- HYDRAZINE TANKS
- ATTITUDE CONTROL NOZZLES
- MINIMAL COMMAND SYSTEM
- EXTENDED COMMAND SYSTEM
- DUAL ATTITUDE CONTROL SYSTEM
- MOTOR
- BATTERIES
- FLYWHEEL
- PRIMARY MIRROR
- FILM HANDLING SYSTEM
- TAKE UP REELS

BIF-008-
BIF008W-C-015446-OH-77

TOP SECRET

BIF008W-C-01546-OH-77

DUAL PLATEN CONFIGURATION

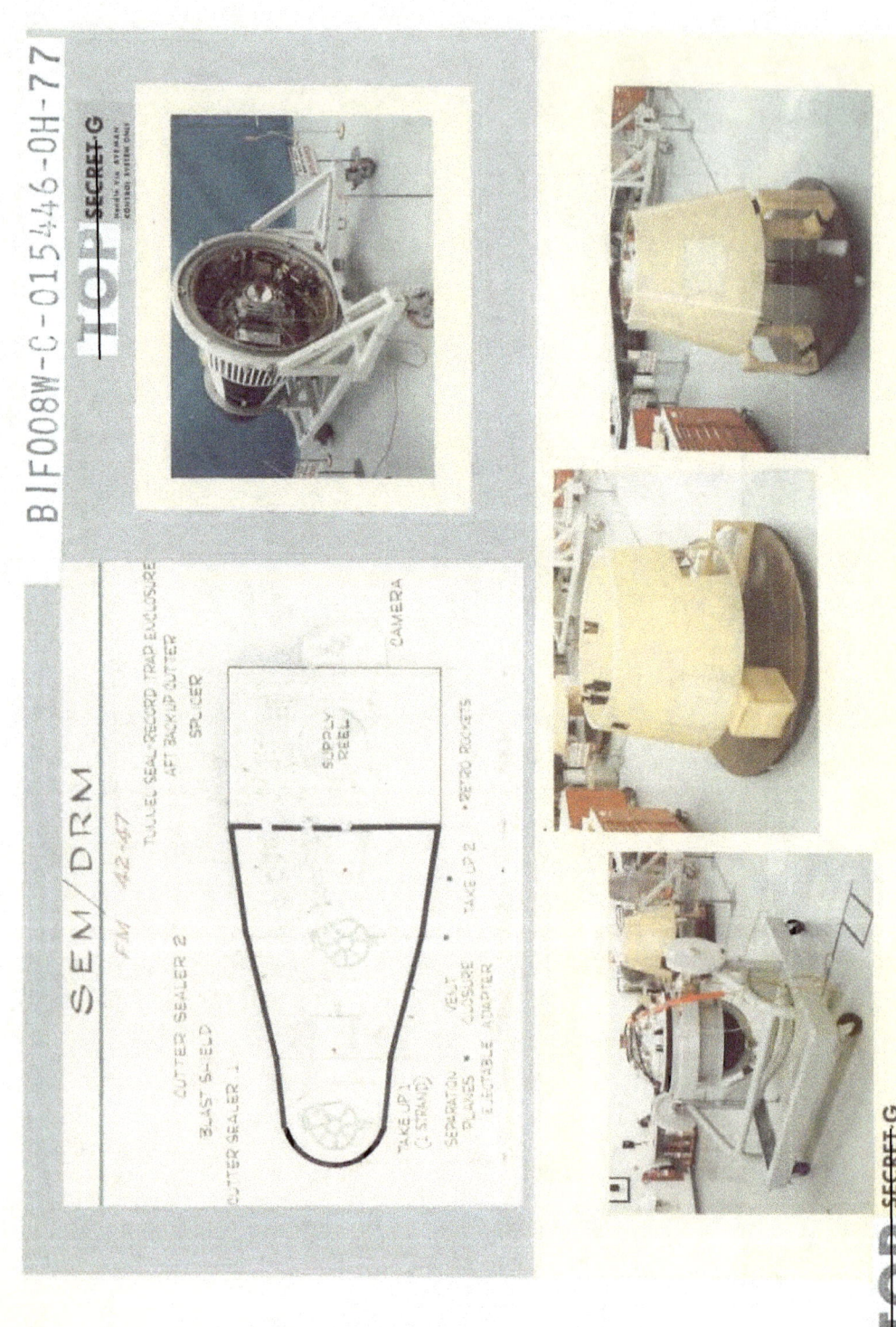

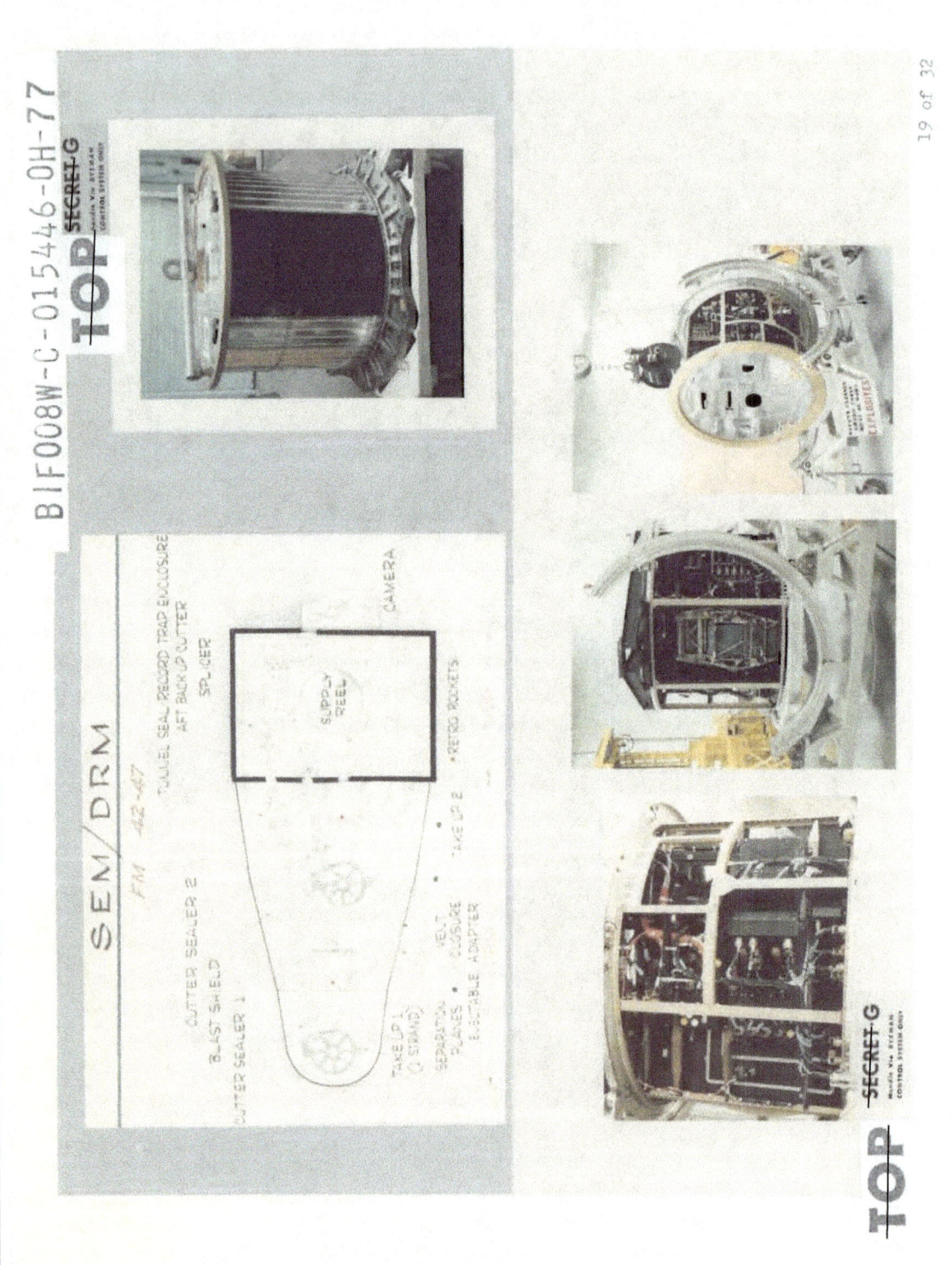

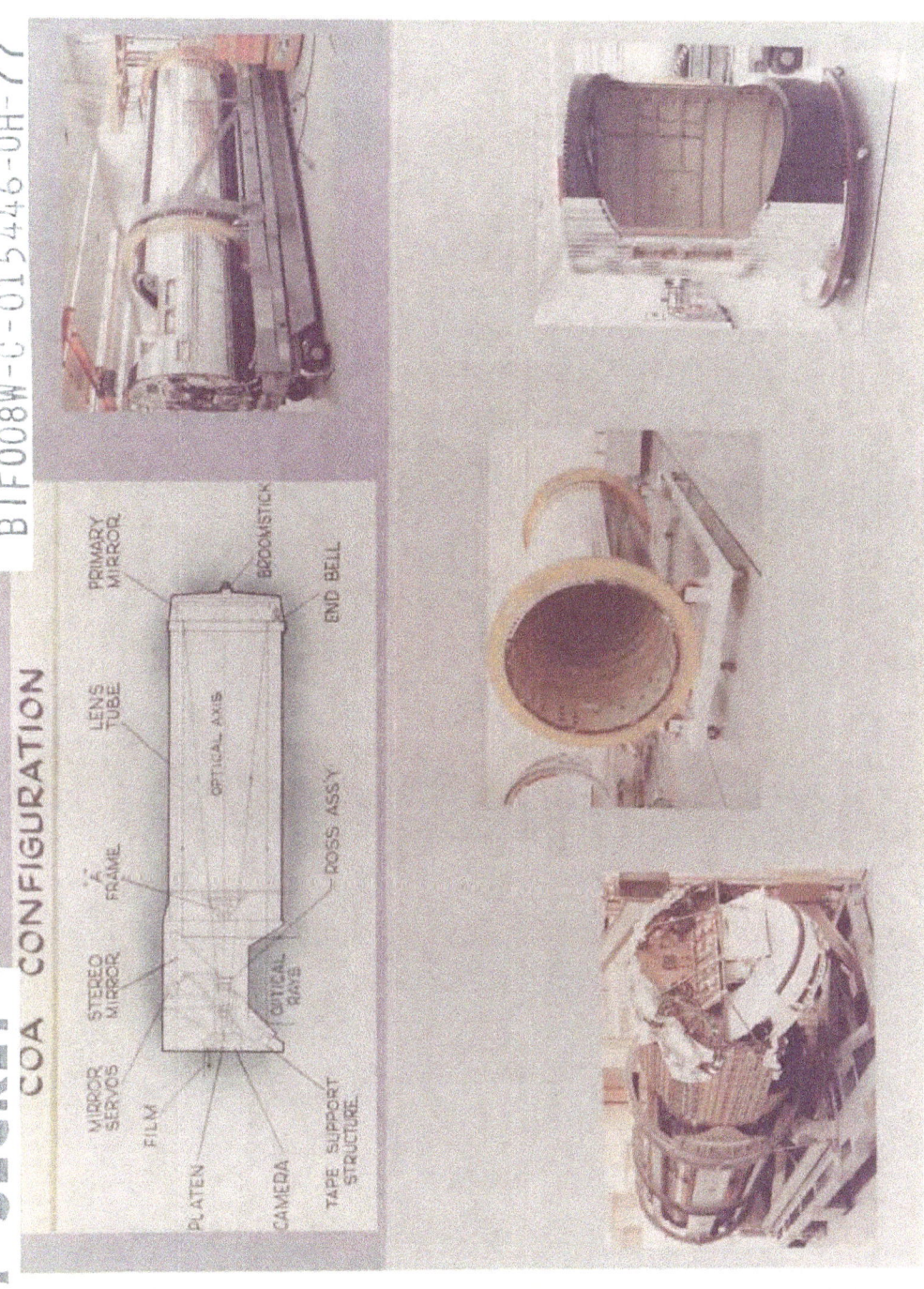

BIF008W-C-015446-OH-77

Location of Program Activities

ROCHESTER — ONF
SUNNYVALE — FAN
PALO ALTO — WCEO
VANDENBERG AFB — FAS
LOS ANGELES — CUSTOMER LIAISON OFFICE
ELMGROVE — H.E.
L.P.

Handle via BYEMAN
Control System Only

Advanced Gambit Orbiting Vehicle

BIF008W-C-015446-OH-77

- STRIP MODE OPERATION

Advanced Gambit
Operational Modes

ROLL

FLIGHT DIRECTION

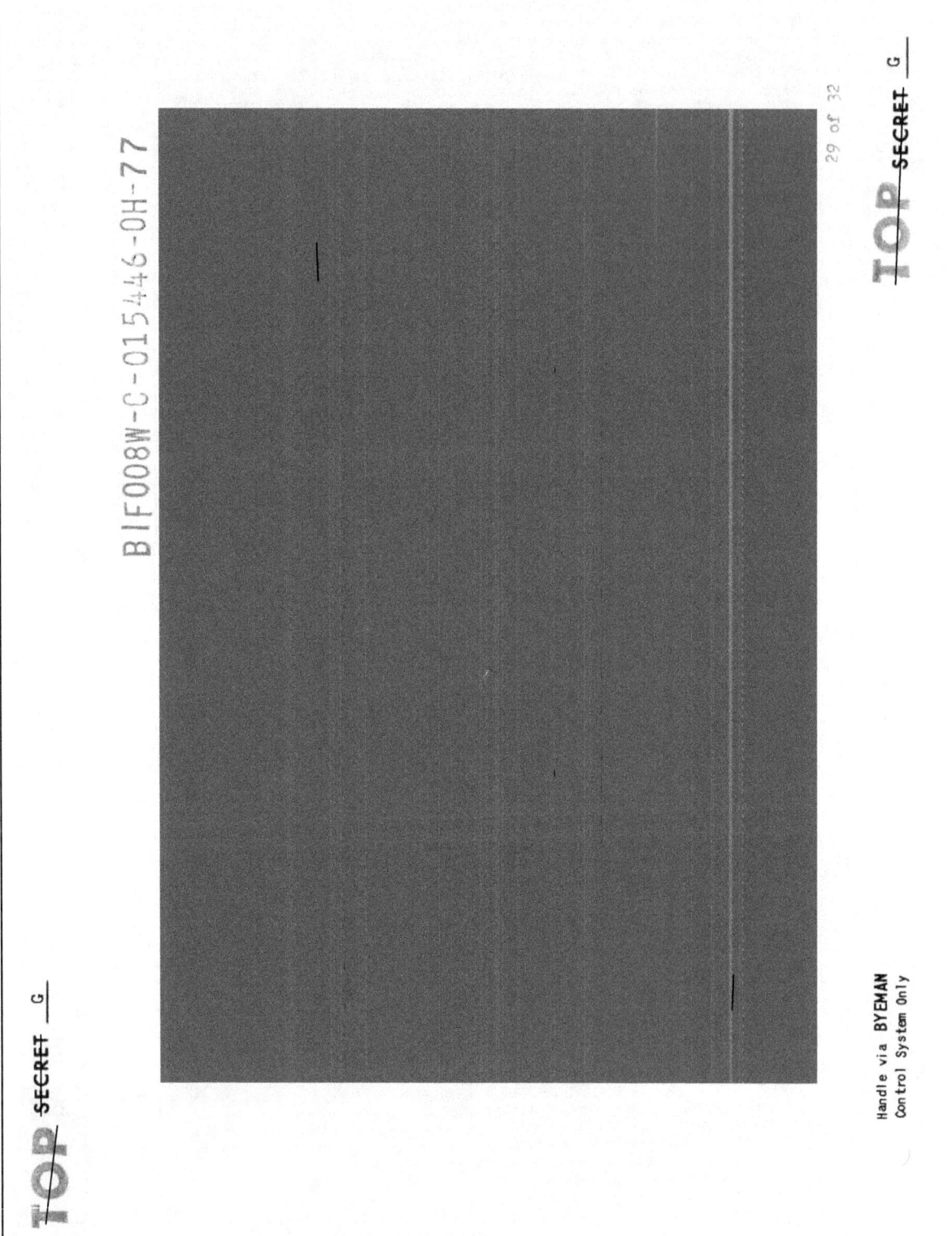

Re-Entry Sequence

DIRECTION OF FLIGHT

- RV-ELECTRICAL DISCONNECT
- MECHANICAL SRV SEPARATION
- SPIN-UP
- RETRO ROCKET FIRE
- SRV DE-SPIN
- THRUST CONE EJECTION
- RECOVERY BEACON
- RV STARTS RE-ENTRY
- CHUTE DEPLOY. INITIATED
- DROGUE CHUTE BLOSSOM. HEAT SHIELD RELEASED.
- MAIN CHUTE REEFED
- MAIN CHUTE BLOSSOM
- AIR RECOVERY
- WATER RECOVERY IF NOT AIR RECOVERY

BIF008W-C-015446-OH-77

Operational Performance Score - History

CONTRACT - 0009 (37 - 47)

OM	%	OM	%
37	100	42	100
38	100	43	100
39	100	44	100
40	100	45	100
41	100	46	100
		47	100

Gambit Dual Mode Capability

BIF008W-C-015446-OH-77

- SURVEILLANCE
  LOW ALTITUDE (75 NM)
  40 DAYS

and

- SEARCH
  HIGH ALTITUDE (500 NM)
  80 DAYS

Center for the Study of National Reconnaissance Classics

GAMBIT (KH-8) IMAGERY
1966 - 1984

CENTER FOR THE STUDY OF
NATIONAL RECONNAISSANCE
CHANTILLY, VA

APRIL 2012

03 AUGUST 1966

**AMMUNITION LOADING & EXPLOSIVES PLANT
KEMEROVO, SOVIET UNION**

POSSIBLE CASTING BUILDING

CASTING BUILDING

CURING BUILDING

CASTING BUILDING

PROPELLANT MIX/BLENDING BUILDING

POSSIBLE CASTING BUILDING

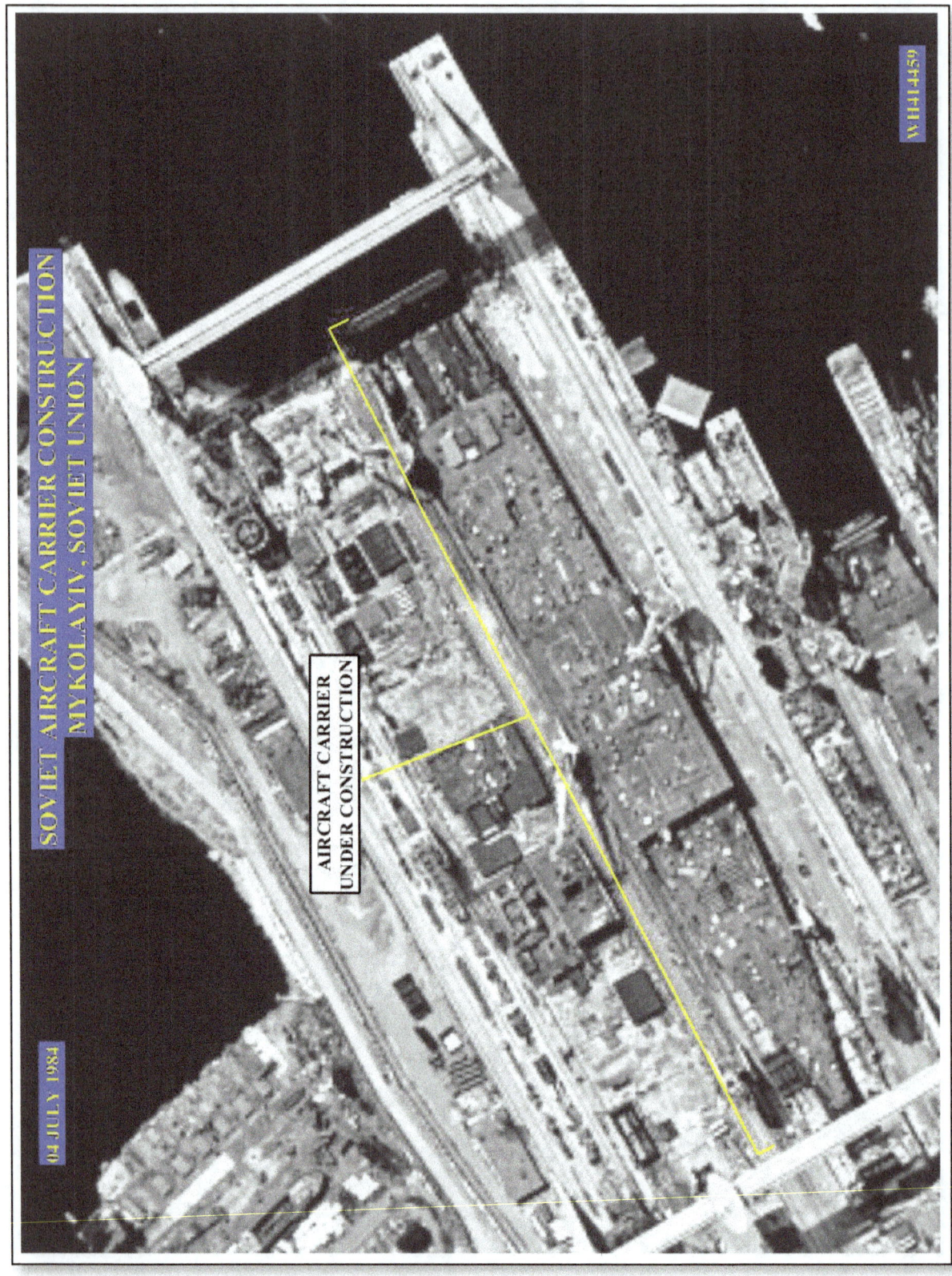

Center for the Study of National Reconnaissance Classics

GAMBIT PROGRAM
YELLOW BRICK ROAD
PRESENTATION

CENTER FOR THE STUDY OF
NATIONAL RECONNAISSANCE
CHANTILLY, VA

APRIL 2012

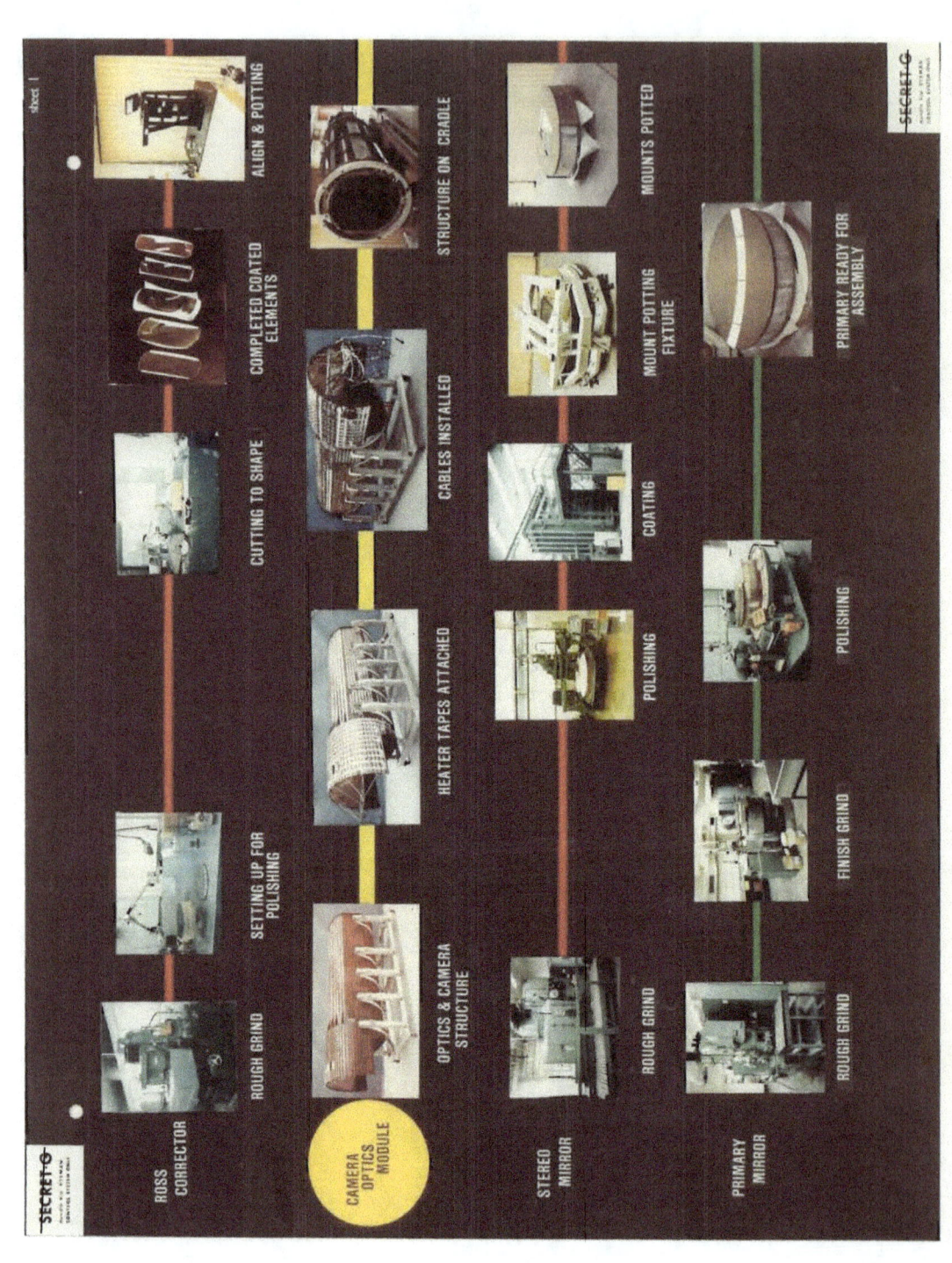

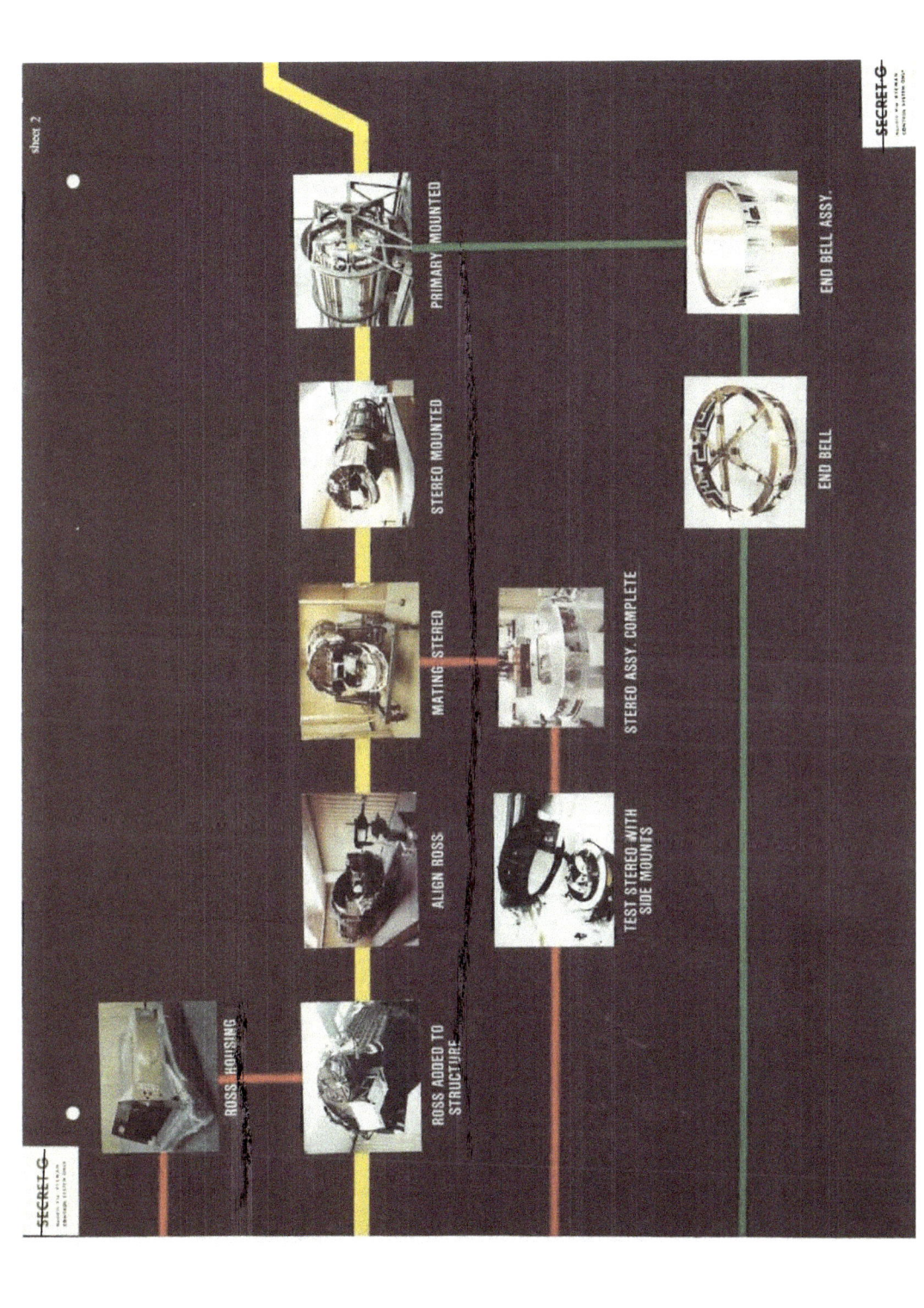

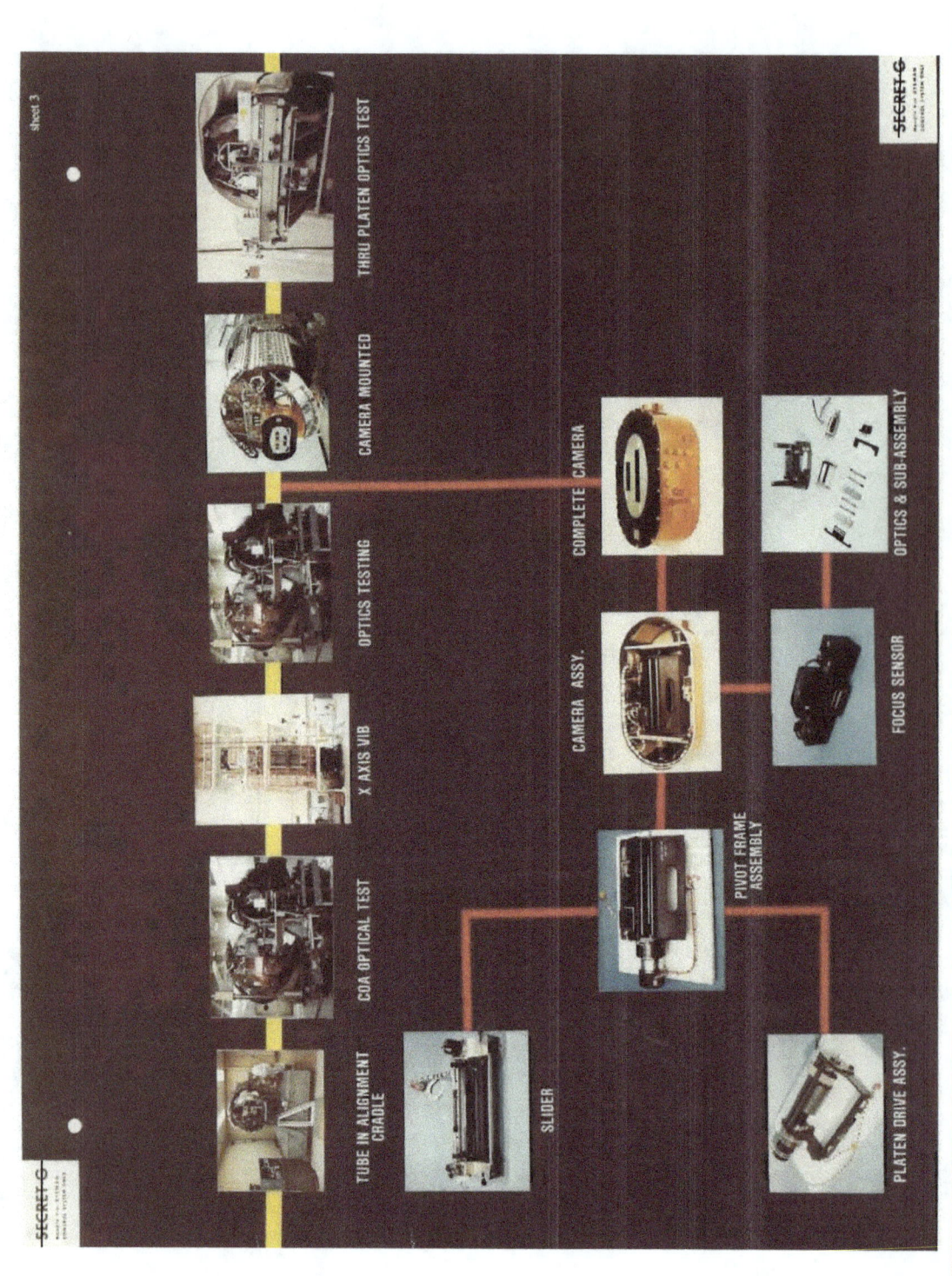

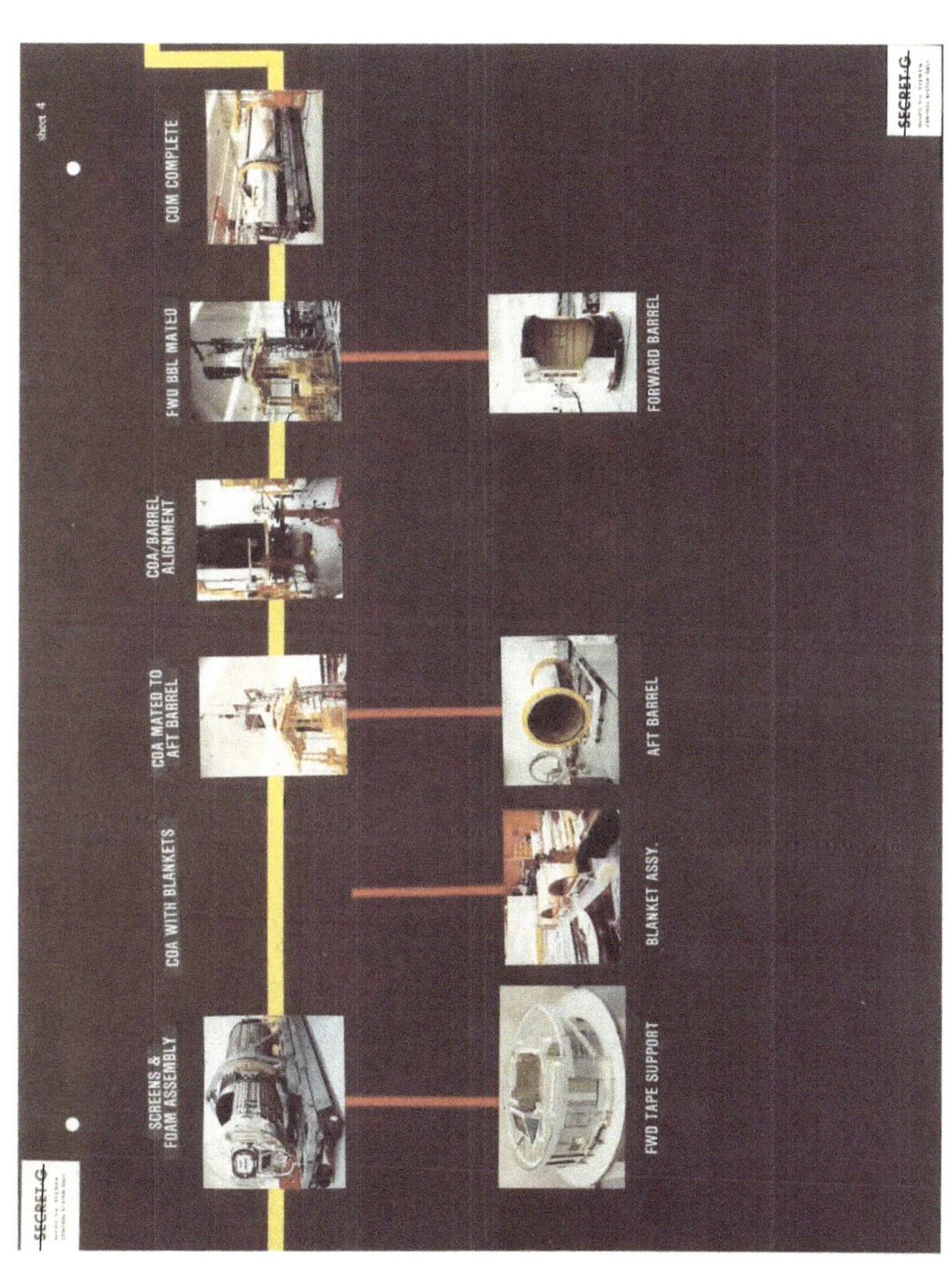

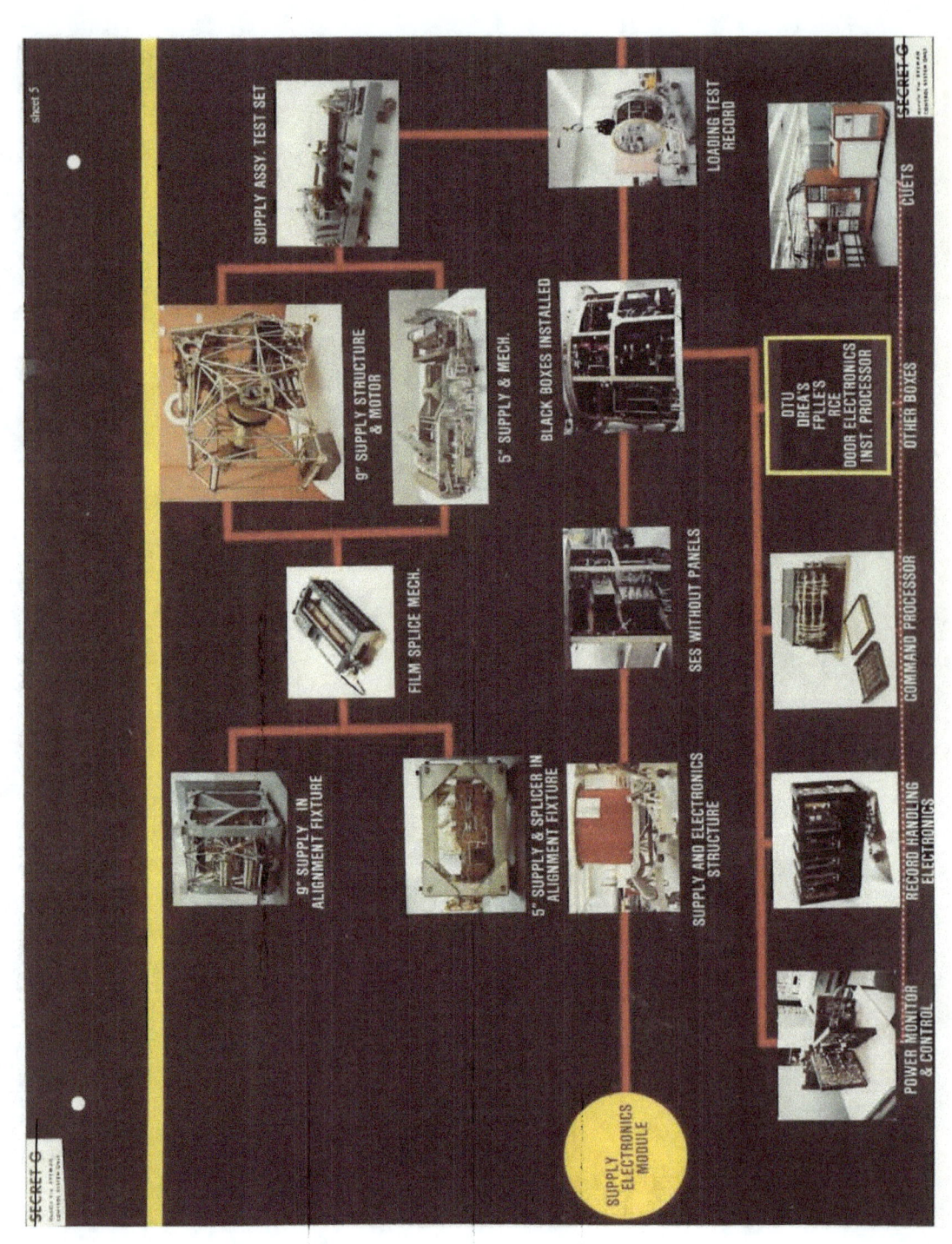

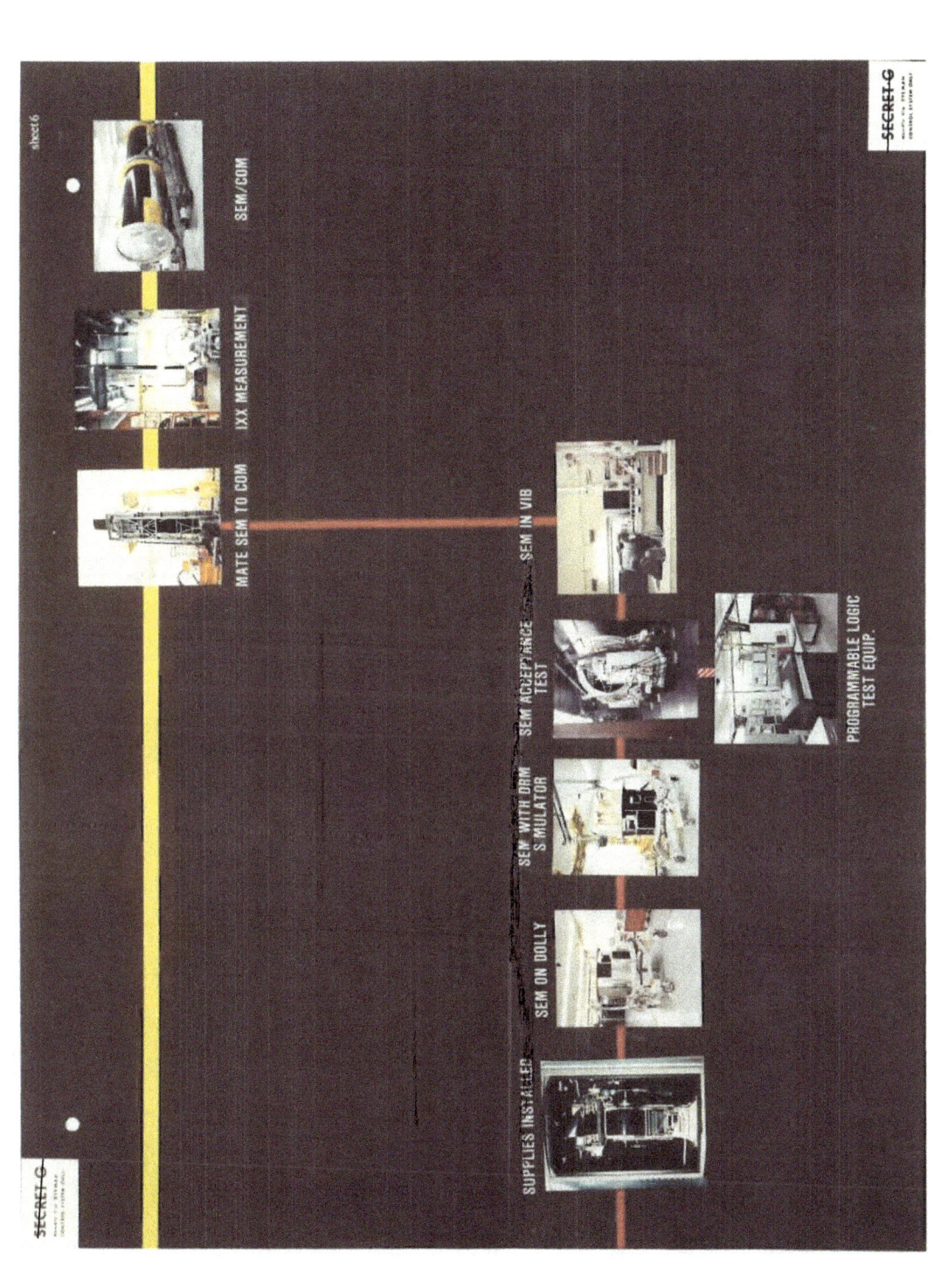

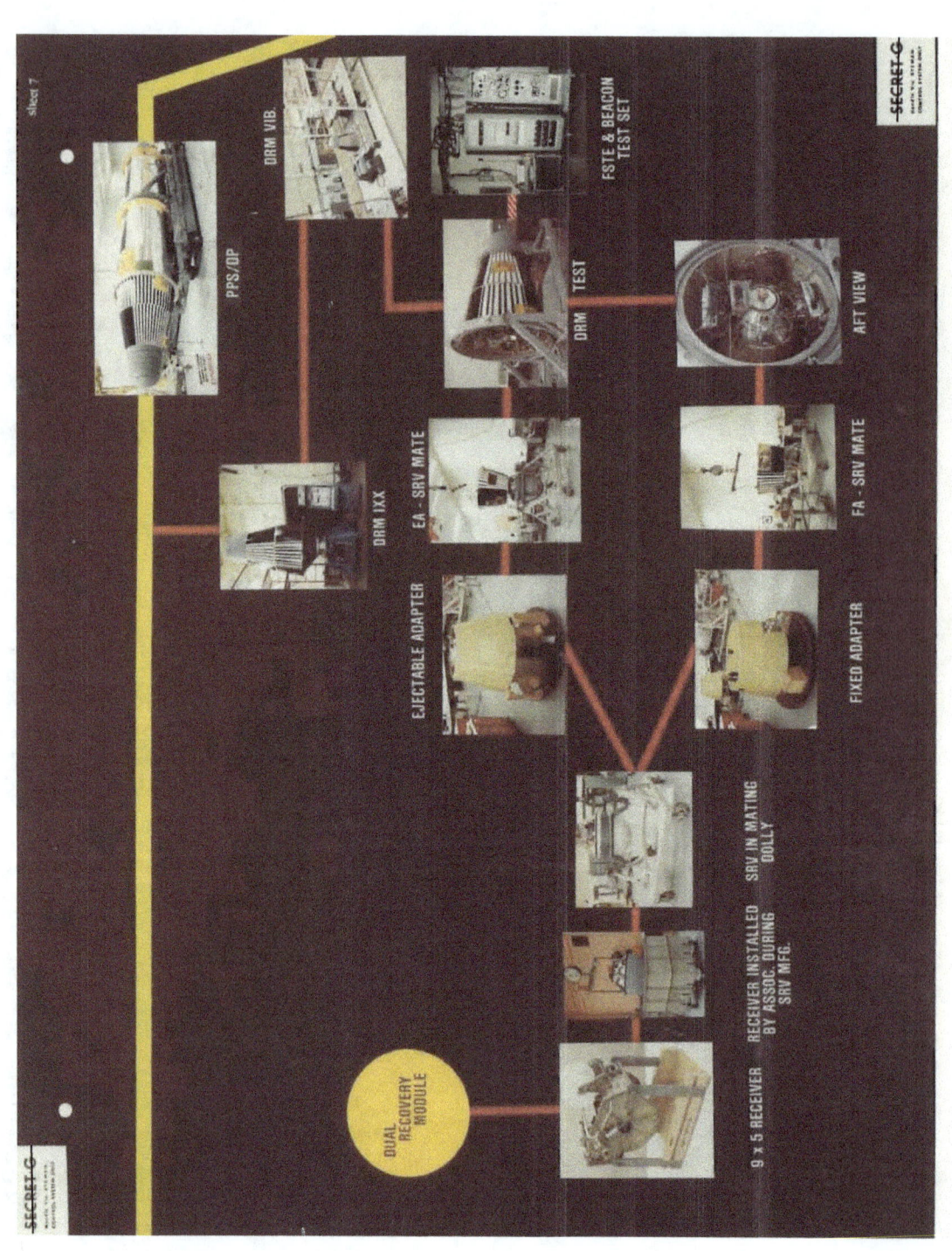

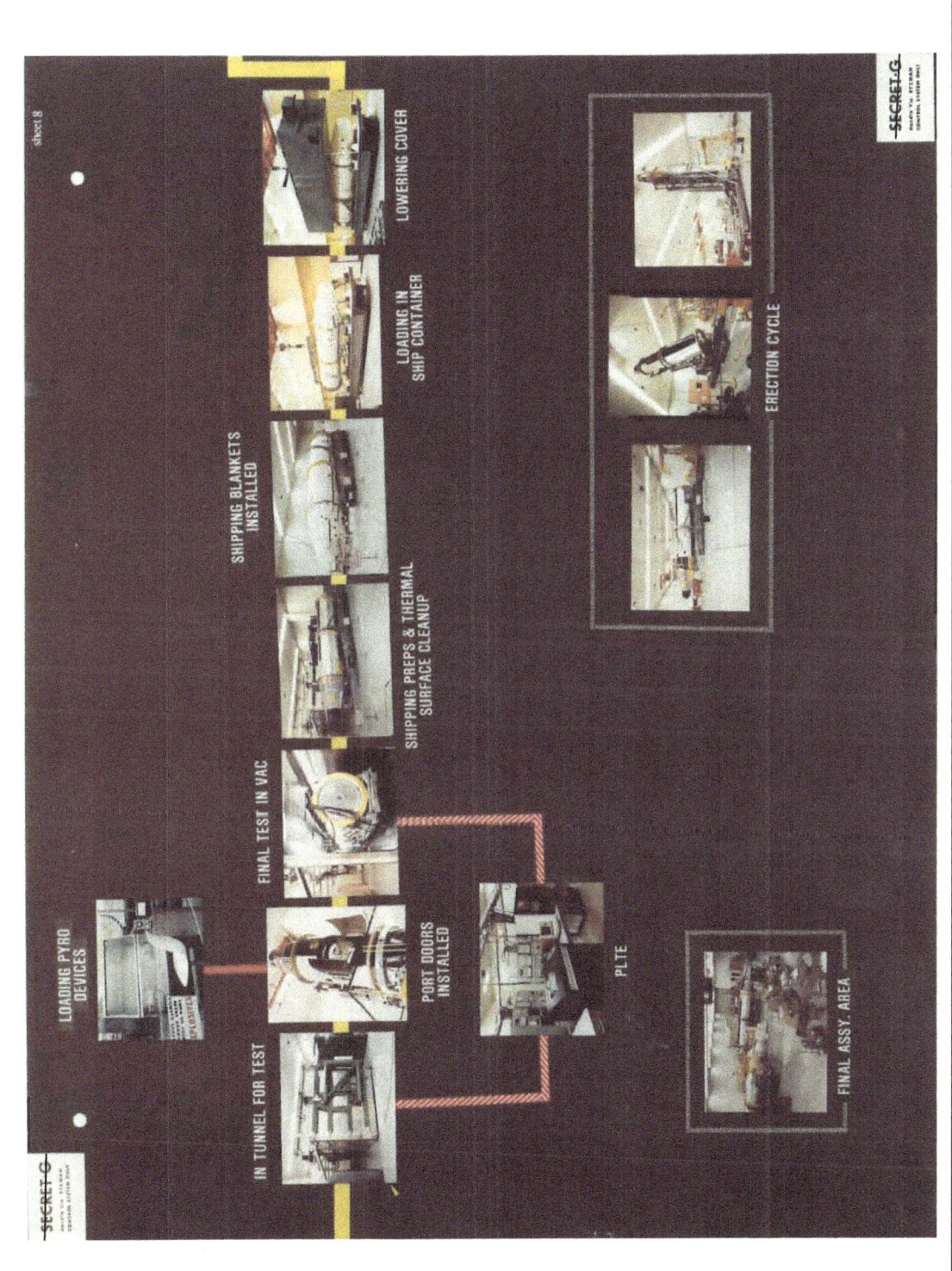

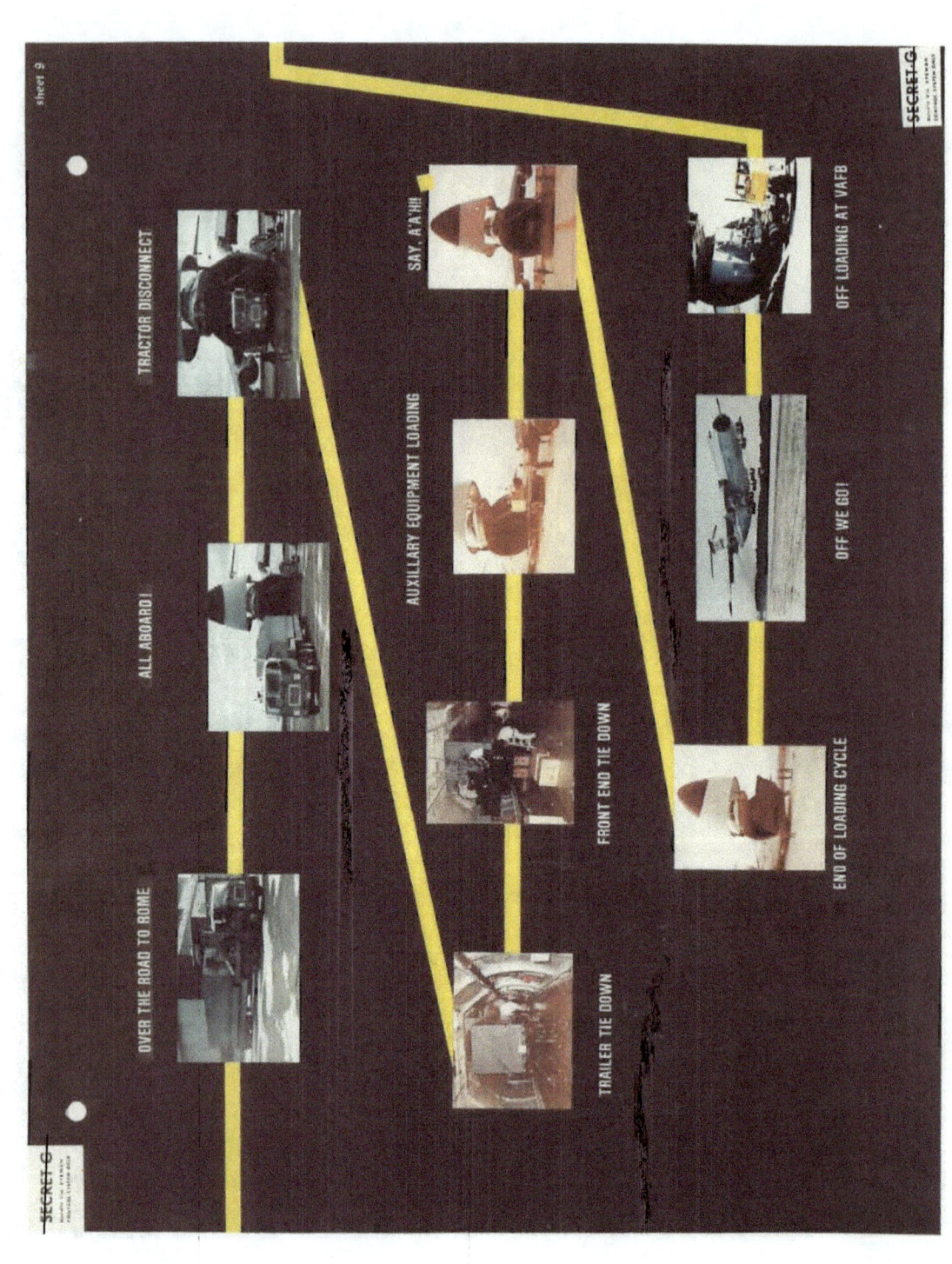

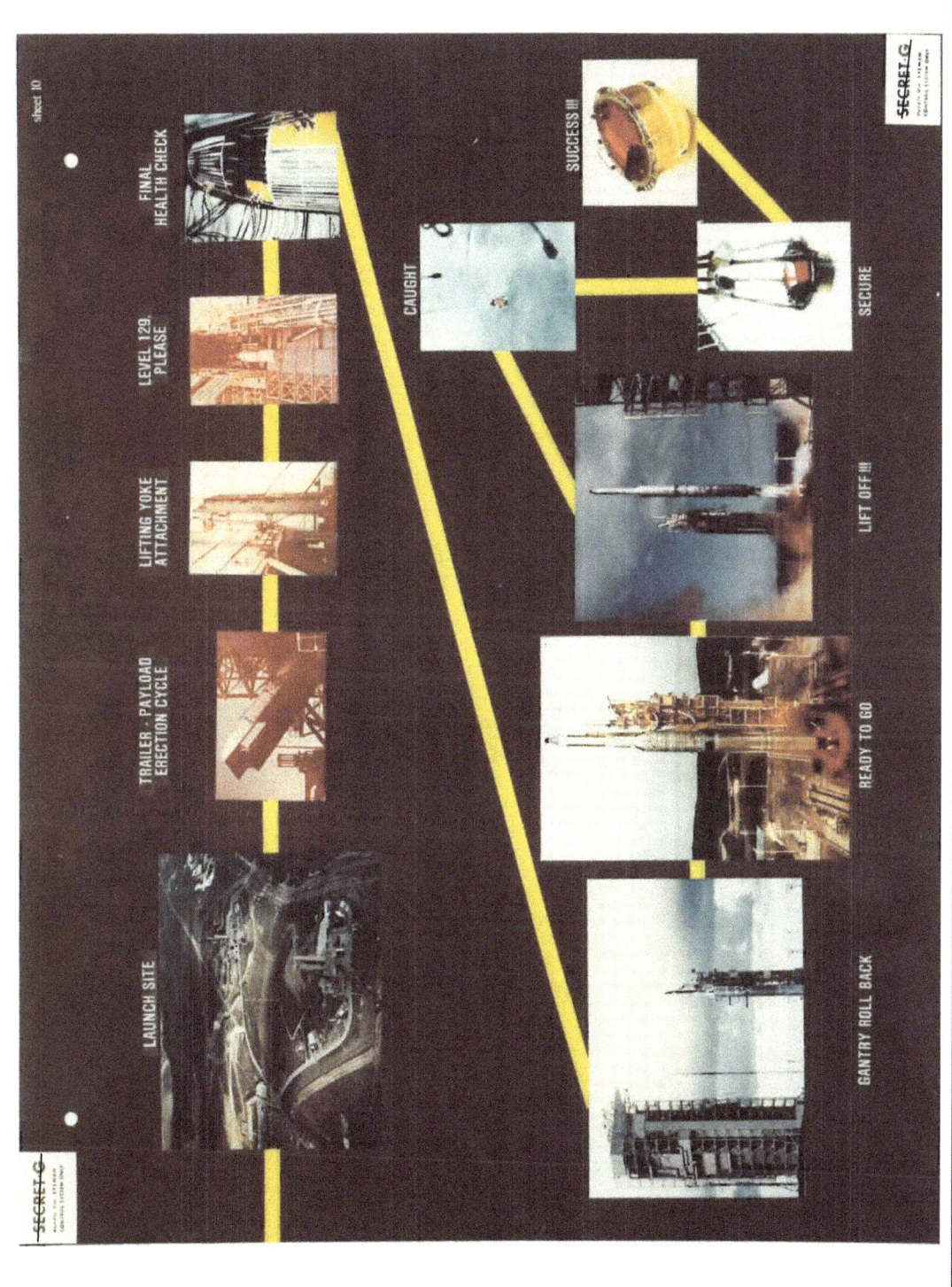

Center for the Study of National Reconnaissance Classics

SECRET

Handle via BYEMAN
Control System Only

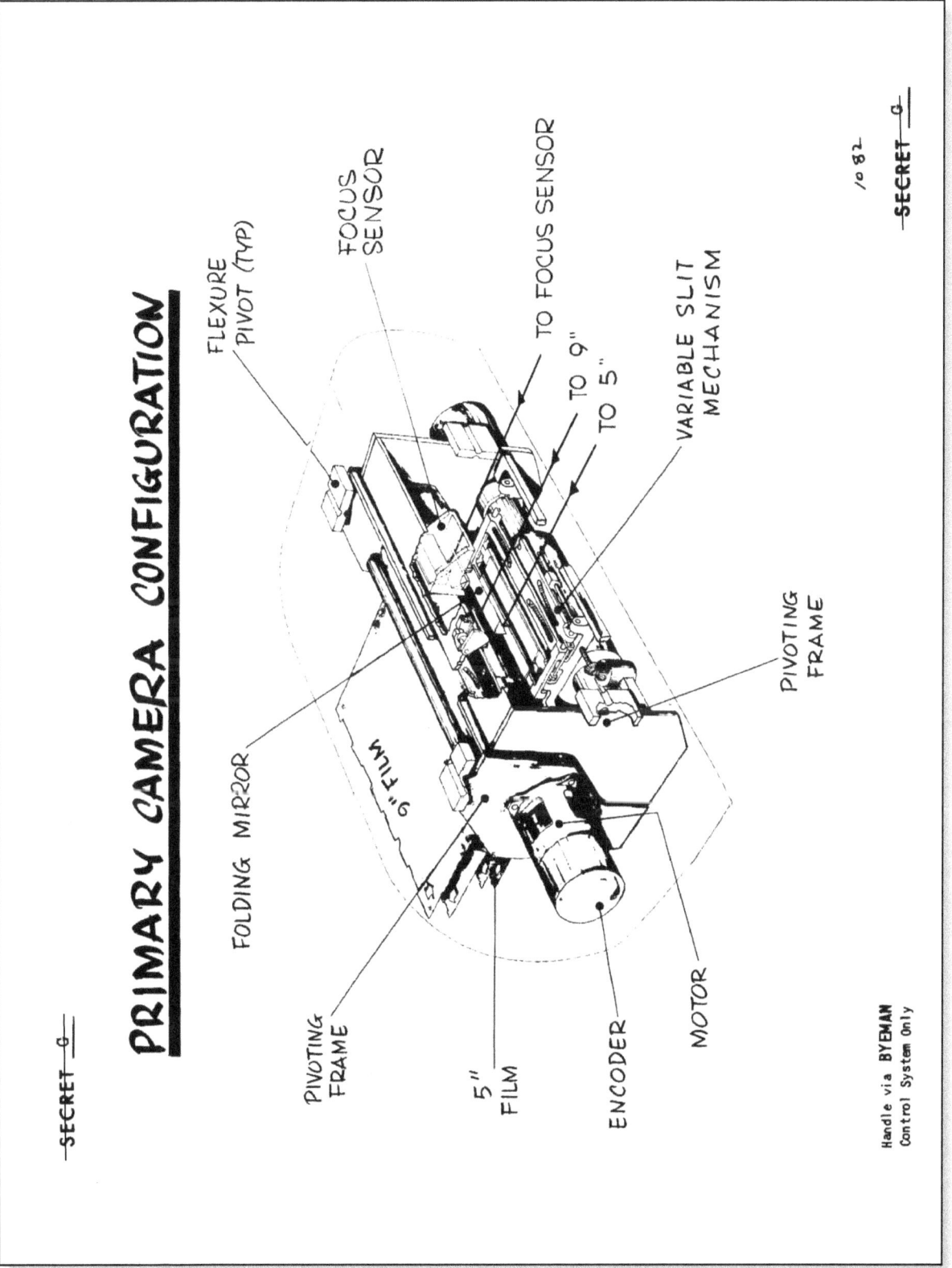